Springer Water

The book series Springer Water comprises a broad portfolio of multi- and interdisciplinary scientific books, aiming at researchers, students, and everyone interested in water-related science. The series includes peer-reviewed monographs, edited volumes, textbooks, and conference proceedings. Its volumes combine all kinds of water-related research areas, such as: the movement, distribution and quality of freshwater; water resources; the quality and pollution of water and its influence on health; the water industry including drinking water, wastewater, and desalination services and technologies; water history; as well as water management and the governmental, political, developmental, and ethical aspects of water.

Kinga Szálkai · Máté Szalai
Editors

Theorizing Transboundary Waters in International Relations

Springer

Editors
Kinga Szálkai
Budapest Metropolitan University
Budapest, Hungary

Máté Szalai
Ca' Foscari University
Marie Skłodowska-Curie Cofund Fellow
Venice, Italy

ISSN 2364-6934 ISSN 2364-8198 (electronic)
Springer Water
ISBN 978-3-031-43375-7 ISBN 978-3-031-43376-4 (eBook)
https://doi.org/10.1007/978-3-031-43376-4

This Springer imprint is published by the registered company Springer Nature Switzerland AG
The registered company address is: Gewerbestrasse 11, 6330 Cham, Switzerland

Paper in this product is recyclable.

Foreword

Questions of conflict and cooperation on transboundary waters have since long spurred interest of scholars in different disciplines. Among these are Geography, Law and Economics, for example, and certainly International Relations, the discipline of which the study of interactions between states is its core.

Theoretical approaches within International Relations (IR) have inspired much of the debate and many case studies on shared waters, with the two major schools of Neo-Realism and Liberal Institutionalism being of most relevance. Miriam Lowi's book "Water and Power" from 1993 is a seminal study of (water) politics in the Jordan basin from a neo-realist perspective. Neo-Realism also was the basis for the hydro-hegemony framework developed by the London Water Research Group (LWRG) that put particular attention to soft forms of power. In a liberal-institutionalist tradition, Claudia Sadoff and David Grey identified the various benefits arising from collaboration on a river, and Erika Weinthal showed in her 2002 book "State Making and Environmental Cooperation" how international actors and side payments made water cooperation beneficial for the newly sovereign states of Central Asia, to name just a few examples. Even if not always explicitly referred to, these two schools of thought provide the backbone of what Frederic Julien in 2012 called the "water security" and the "water rationality" discourses—two discourses stressing either the security risks stemming from shared water resources or their cooperative potential.

However, more often than not, the reference to IR theories in studies on international waters has been implicit and their understanding and conceptualization rather superficial—a fact criticized early on. Scholars also have identified inherent weaknesses of mainstream IR approaches for understanding hydropolitics, e.g. their state centredness and their positivist-rationalist ontology. This shortcoming is not necessarily grounded in IR theory as such, but in a rather selective reading of it by many water scholars. In addition, developments in a few prominent, conflict-prone basins—like the Nile, Jordan, and Indus—have informed much of the academic discourse, biasing analysis towards theorization focussed on conflict and unequal power.

Consequently, water scholars have broadened their theoretical tools and lenses. They started to make use of constructivist approaches in IR and furthered conceptualizations of the role of non-state actors, discourses, and questions of identity in conflict and cooperation around transboundary waters. This paved the way for a more nuanced and sophisticated understanding of transboundary dynamics beyond the "black or white" understanding of conflict vs cooperation.

Against this backdrop, the present volume, "Theorizing Transboundary Waters in International Relations" is a much-needed contribution to the academic debate. It puts IR theory and concepts in the centre, uses its more critical approaches, and brings in global perspectives from a variety of regions and basins. Chapters in this book include the analysis of the role of non-state and subnational actors, assessment of the explanatory value of different IR theories, and discussions of the value of concepts like discourse, trust, and water security for understanding transboundary water interactions. The various case studies show how a conscious application of theoretical concepts can help to better understand water diplomacy processes or key issues like data sharing and public health challenges.

My wish for this book is that it stimulates and triggers further engagement of the multi-disciplinary scholarly community on transboundary waters with theories of International Relations. The book opens up a space in which much more approaches are still to discover and to be used in analysis of transboundary water interactions. This would not only improve our understanding of transboundary waters better with the means of IR; it could also make the study of international waters, a field to better understand the overall dynamics of International Relations.

Jenniver Sehring
Associate Professor of Water
Governance and Diplomacy
IHE Delft
Delft, The Netherlands

Contents

Contributors

Atal Ahmadzai St. Lawrence University, Canton, NY, USA

Sameh W. H. Al-Muqdadi Green Charter, Malsch, Germany

Ali Oguz Dirioz International Entrepreneurship Department, TOBB University of Economics and Technology, Ankara, Turkey

Mary Durfee Michigan Technological University, Houghton, MI, USA

Gabriel Eckstein Texas A&M University School of Law, College Station, TX, USA

Fatine Ezbakhe Institute for Environmental Sciences (ISE), UNESCO Chair in Hydropolitics, Geneva Water Hub, University of Geneva, Geneva, Switzerland

Seyedeh Zahra Ghoreishi Department of Irrigation and Reclamation Engineering, Faculty of Agricultural Engineering and Technology, College of Agriculture and Natural Resources, University of Tehran, Karaj, Alborz, Iran

Milad Jafari Department of Water Engineering and Management, Faculty of Agriculture, Tarbiat Modares University, Vardavard, Tehran, Iran

Ayşegül Kibaroğlu Department of Political Science and International Relations, Faculty of Economics, Administrative and Social Sciences, MEF University, İstanbul, Turkey

Najibullah Loodin Water Management and Hydrological Science Program, College of Arts and Sciences, Texas A&M University, College Station, TX, USA

Péter Marton Corvinus University of Budapest, Budapest, Hungary

Hojjat Mianabadi Department of Water Engineering and Management, Faculty of Agriculture, Tarbiat Modares University, Vardavard, Tehran, Iran

Rosario Sanchez Texas Water Resources Institute, Texas A&M University, College Station, TX, USA

Bota Sharipova IHE Delft Institute for Water Education, Delft, The Netherlands

Vijay P. Singh Department of Biological and Agricultural Engineering and Zachry Department of Civil and Environmental Engineering, Texas A&M University, College Station, TX, USA

Máté Szalai Ca' Foscari University, Venice, Italy

Kinga Szálkai Budapest Metropolitan University (METU), Budapest, Hungary

Zoltán Vörös University of Pécs, Pécs, Hungary

Chapter 1
Introduction: Mixing Water and International Relations Theory

Kinga Szálkai and Máté Szalai

Abstract 15 years ago, a discussion started in Political Geography about the role of International Relations theories in understanding transboundary water issues. Furlong drew attention to the fact that the application of IR theories in transboundary water analysis is "implicit and unacknowledged", which means a significant constraint for research programs. Zeitoun and Warner also admitted that the importance of the explicit integration of political theories "cannot be said to be truly resonant yet" in the global water community. Since then, the connections between IR theories and transboundary water analysis have been problematized more intensely, and explicitly theoretical accounts on transboundary waters are being published from time to time. The aim of this Introduction is to prepare the ground for the collection of the latest state-of-the-art findings of scholars working on related issues with reviewing the main traditional approaches towards theorizing transboundary waters within IR.

Keywords International Relations · Theory · Transboundary waters · Neorealism · Neoliberalism

1.1 Introduction

As water resources are globally under growing pressure, the importance of mitigating water-related tensions and managing transboundary waters is also increasing significantly. The main aim of this book is to provide a complex overview of how the application of IR-related theoretical frameworks contributes to the understanding

K. Szálkai (✉)
Budapest Metropolitan University (METU), Budapest, Hungary
e-mail: kszalkai@metropolitan.hu

M. Szalai
Ca' Foscari University, Marie Skłodowska-Curie Cofund Fellow, Venice, Italy
e-mail: mate.szalai@unive.it

K. Szálkai and M. Szalai (eds.), *Theorizing Transboundary Waters in International Relations*, Springer Water, https://doi.org/10.1007/978-3-031-43376-4_1

and the management of transboundary water relations, how the different perspectives and frameworks may support the participation and the cooperation of various stakeholders, and what kind of further added value they may offer.

15 years ago, a discussion started in Political Geography (Furlong 2006; Warner and Zeitoun 2008). Furlong drew attention to the fact that the application of IR theories in transboundary water analysis is "implicit and unacknowledged", which means a significant constraint for research programs. Zeitoun and Warner also admitted that the importance of the explicit integration of political theories "cannot be said to be truly resonant yet" in the global water community. Since then, the connections between IR theories and transboundary water analysis have been problematized more intensely, and explicitly theoretical accounts on transboundary waters are being published from time to time. IR theories offer a diverse range of possible frameworks for analysing transboundary waters. Several scholars have been working on the utilization of grand theories such as Realism, Liberalism, and Constructivism; applied critical lenses; or used the various ideas of hegemony theories, de/securitization, theories of water diplomacy and conflict resolution, discursive approaches, or governance theories to discuss and understand the complexity of relations over transboundary waters. Nevertheless, practical approaches and implicit theoretical assumptions are still in overwhelming majority in the related academic works, and there is also no clear consensus on the added value of theorizing transboundary water relations among the practitioners of the field.

The overarching purpose of this book is to contribute to the literature on the explicit theorization of transboundary water issues within the discipline of International Relations. The main objectives of it are to contrast the traditional and the current theoretical approaches towards understanding the general dynamics of transboundary water conflict and cooperation; to collect the latest state-of-the-art findings of scholars working on the theorization of transboundary water relations; and to demonstrate that new and transformative perspectives can arise in the intersections of theory and practice, which can have a significant added value not only for academics, but also for practitioners and other stakeholders, while fostering the discussion in the global water community. Through the analysed case studies, the volume also intends to provide an overview of significant current hotspots and best practices of water management issues. Comprising all these features, this book is a great companion for both academics and practitioners dealing with transboundary water issues in any phases of their careers.

1.2 Transboundary Water Politics and International Relations

"The statement that 'water is politics' hardly needs any defense"—Peter P. Mollinga claimed in his 2001 paper, in response to comments that water issues are becoming increasingly important in International Relations (Mollinga 2001: 733). The role of transboundary waters in the development of interstate conflicts had been an important topic of International Relations theory at that time for decades already, similarly to the question of how shared water resources affect cooperation between countries. According to the most traditional definition of Convention on the Protection and Use of Transboundary Watercourses and International Lakes, ""transboundary waters" means any surface or ground waters which mark, cross or are located on boundaries between two or more States." As there are 263 transboundary lake and river basins shared by 145 states with various relations (UN Water 2008: 1), the role of conflict and cooperation over them indeed cannot be ignored in the discipline of International Relations.

"The systematic study of interstate conflict and cooperation over transboundary water issues" (Elhance 1999: 3) emerged in the late 1970s (LeMarquand 1977), traditionally as an empirically oriented approach. As a consequence of this, many detailed case studies have been published on transboundary waters, and practical examples of water-related tensions, water sharing and water governance have been popular topics among experts. In the meantime, theoretically-oriented research has been significantly less widespread. The main reason for this is that the first experts on transboundary waters were not primarily political scientists or specialists in International Relations, but rather geographers, agricultural experts, engineers or lawyers, who were either not familiar with, or did not feel the need to apply the frameworks of International Relations theory on their studied cases (Warner and Zeitoun 2008: 802). In addition, transboundary water experts coming from the fields of political and social sciences neither emphasised the importance of the conscious application of theories, and the most well-known authors, such as Peter H. Gleick or Thomas F. Homer-Dixon, did not explicitly link their research to any of the existing IR theoretical frameworks (du Plessis 2000: 10). It should also be pointed out, however, that the lack of explicit theoretical contributions does not mean that the theories of International Relations have completely been excluded from the analysis of transboundary waters. Instead, they often appear in a 'hidden' or implicit form, unconsciously structuring the reality perceptions of experts and their readers alike, especially in relation to approaches to conflict and cooperation (Furlong 2006: 439).

Already Furlong, Warner and Zeitoun observed a kind of reversal in this trend before the end of the 2000s, and pointed out that the number of experts who consciously and explicitly apply theoretical frameworks is growing. Nevertheless, at that time, most of the theoretical contributions were constrained to the approaches of grand theoretical schools, such as neorealism and neoliberalism, and the number of academic articles focusing on non-rationalist or critical approaches was negligible (Warner and Zeitoun 2008: 803). In the past 15 years, the tendency of paying more attention to explicit IR theoretical frameworks in the study of transboundary waters continued. In the meantime, in line with the general direction of development in International Relations theory, the application of non-classical approaches and critical lenses also strengthened.

In this book we start out from the assumption that the conscious application of the frameworks of International Relations theory help scientists, experts, policy- and decision-makers, and further various stakeholders alike to recognize the patterns they use to make sense of transboundary water relations beyond everyday practicalities. These patterns influence the factors which various actors take into consideration, their general assumptions about the nature of conflict and cooperation or the main functions and priorities of water use, and the kind of tools that they perceive to be efficient to tackle the emerging issues. Thus, they play an important role in forming transboundary water politics, and the explicit reflections of them may contribute to the understanding and the management of transboundary water relations significantly.

1.3 Neorealism and Neoliberalism: Grand Theoretical Schools and Transboundary Waters

As Furlong, Zeitoun and Warner observed, the so-called "grand theories" of International Relations provided the basis for the traditional implicit way of theorization in the studies on transboundary waters, and they were the first to surface as the application of explicit approaches became more and more common. Besides, the assumptions and the patterns of these theories are often used—consciously or unconsciously—in the non-academic fields of journalism, politics, or diplomacy discussing transboundary water issues, especially in relation to conflict and cooperation. The two mainstream schools of International Relations, neorealism and neoliberalism are often used to provide and inspire explanations for them. Therefore, before engaging with the contemporary research initiatives on theorizing transboundary waters, it is worth to review the central ideas of these two schools in a nutshell, applied to transboundary waters as the Neo-Malthusian and the Cornucopian approaches.

The Neo-Malthusian approach (e.g. Gleick 1993, 2010; Homer-Dixon 1994, 1999) is named after the British author Thomas Malthus. In his *Essay on the Principle of Population* of 1798, he was the first to formulate the thesis that the scarcity of resources, especially if it is made more predominant by population growth, can lead to famine or war. The followers of Neo-Malthusianism in the twenty-first century are generally concerned with conflicts that may arise from the growing gap between the increasing demand for finite resources and the ever-shrinking supply due to overuse, demography, or environmental factors (Kipping 2009: 712). Neo-Malthusian theory is linked to transboundary waters through the assumption that resources, including water, are scarce goods, the supplies of which become more limited with population growth and environmental change. According to them, this is likely to lead to increasing tensions, as in Malthus' original theory (Malthus 1798/1998). As Thomas Naff summarized almost 200 years later: "the strategic reality of water is that under circumstances of scarcity, it becomes a highly symbolic, contagious, aggregated, intense, salient, complicated, zero-sum, power-and prestige-packed issue, highly prone to conflict and extremely difficult to resolve" (Naff 1994: 274).

The finite nature of water resources does not only raise the threat of conflict from the aspect of water scarcity, but also means high vulnerability for states through the interdependencies related to the possession of transboundary waters. Water, as a strategic resource, is essential for states' security and survival, thus, theoretically, if possible they strive for absolute and exclusive sovereignty over it. In the case of transboundary waters, this cannot be exercised, and therefore states must interact with each other to ensure water security and water supply. According to the neo-Malthusian approach, this situation is likely to lead to rivalry for water, which can motivate one-sided state actions to defend resources and interests, creating a constant source of tensions. The theory therefore claims that there is a high potential for conflict between states sharing transboundary waters.

Neo-Malthusian claims have been widely refuted by experts who criticize their narrow focus on water conflict. They argue that Neo-Malthusians tend to ignore distinctions between water disputes, conflicts, tensions and water wars, which can be misleading in their argumentations. In the meantime, Neo-Malthusians are also often accused with ignoring the differences between specific contexts surrounding transboundary water issues, which may have a similar effect. For example, factors like population size, climate, access to water, intensity of trade relations, or the level of technological development each have an important role in assessing water scarcity. Furthermore, statistics on transboundary water conflict and cooperation indicate (Wolf 2013) that Neo-Malthusians tend to concentrate on extreme cases to indicate the threats and the challenges of water sharing, while the vast majority of transboundary water relations have been manageable throughout human history, and cases of cooperation are much more frequent than those of conflict (Wolf et al. 2003: 31).

The neoliberal Cornucopian approach, which provides the most prominent mainstream critique of the neorealist Neo-Malthusian approach, does not deny the idea that the political, economic, environmental and geographical interdependencies arising in transboundary water relations have the potential to lead to conflict between the concerned states. Nevertheless, their expectations about the related challenges point to the opposite direction, as they assume that these interdependencies push states towards finding common solutions, thus fostering cooperation instead of conflict. The Cornucopian approach assumes that the sum of the relative benefits that can be gained from cooperation exceeds the benefits that can be gained through unilateral steps inciting tensions. In the extreme case of an armed water conflict to gain exclusive sovereignty over resources, the probability of victory is often difficult to predict, while the short- and long-term costs and the direct material, economic and political impacts take their toll both on the victorious and the defeated parties. It is therefore more cost-effective for states to avoid conflict, not only directly in terms of water wars, but also in terms of escalating tensions. Cornucopians therefore believe that even if our water resources are finite, and transboundary water relations need to be carefully managed to mitigate tensions, interdependencies are more likely to point towards cooperation.

Cornucopians identify several alternative approaches for reducing tensions over transboundary waters and increase the benefits of cooperation. For example, the theory of environmental peacebuilding applied to transboundary water relations reverses the central argument of Neo-Malthusianism. Instead of assuming that scarcity leads to conflict, the theory of environmental peacebuilding explores the possibilities of how the common use of transboundary waters can facilitate wider cooperation between riparians, creating spill-overs to other fields of policy and fostering closer relations between the concerned states (Conca and Dabelko 2002). The most widely applied Cornucopian approach to mitigate tensions over transboundary waters is the one of liberal institutionalism, which, in this case, refers to establishment of water regimes or specific organizations of water governance, emerging on the global, the regional, or the local level, or in the form of basin-level cooperations. In the words of Helga Haftendorn, a water regime is established when "the affected states to a conflict observe a set of rules designed to reduce conflict caused by use, pollution or division of a water resource or the reduction of the standing costs and the observance over time of these rules" (Haftendorn 2000: 65). In addition to legal and institutional solutions, Cornucopians also claim that technological development may play a major role in mitigating transboundary water conflict. For example, the conditions of joint water management can be improved with new technologies of information sharing, and the development of water-related infrastructures increases the efficiency of the use of water resources. Another related approach is to increase water resources available for human use or consumption, and to reduce pressure on water resources with new technologies.

The authors of this edited volume from time to time refer to the approaches of traditional "grand theories" of International Relations, in order to develop them further, to provide critiques, or to review the roots of the literature related to their research. Their work reveals the complexity and the variety of the explicit application of IR theoretical approaches on transboundary water issues beyond neorealist Neo-Malthusian and neoliberal Cornucopian ideas. They introduce new and transformative perspectives that arise at the intersections of theory and practice to contribute to the more comprehensive understanding of transboundary water politics.

1.4 About This Volume

Overall, the work intends to demonstrate that the theorization of transboundary waters can significantly contribute to the deeper understanding and the more efficient management of water conflicts and cooperation from several aspects. The authors come from diverse backgrounds, and their individual careers are often related to the intersections of theory and practice in the field of transboundary water management. Their expertise covers water issues from all around the globe, which is reflected in the range of the analyzed case studies. The diversity of the experts involved, their backgrounds, their perspectives, the applied theories, and the analyzed cases was an important priority for the editors.

The discussion of their findings is divided into four main sections. The first section deals with reflections and critiques on the grand theories of International Relations, proposing new and more nuanced frameworks for understanding and managing transboundary water relations by going beyond the traditional assumptions. *Sameh Al-Muqdadi* proposes the concept of the hydropolitical cycle (HPC) to connect various theoretical approaches into one, complex explanatory framework applicable to different stages of transboundary water conflict, using the case study of the Euphrates and Tigris River Basin to support his claims. *Atal Ahmadzai* reviews the existing literature of water conflict and cooperation to criticize traditional approaches towards the role of the state, then explores the sub-state–supra-state nexus in transboundary water politics in the case of Pakistan. The third chapter in this part focuses on trust in transboundary water cooperation: *Bota Sharipova* applies a constructivist framework for understanding the role of this traditionally rarely examined factor, anchoring her discussion in the Aral Sea Basin in Central Asia as a case study.

The second section focuses on the catalysts and barriers of cooperation, applying theoretical frameworks which reveal the consequences of the dynamics in power relations and institutional frameworks. Chapter *Zoltán Vörös* uses the theory of hydro-hegemony to explore the position and the interests of China as the obvious hegemon in the Lancang/Mekong River Basin. *Seyedeh Zahra Ghoreishi*, *Hojjat Mianabadi*, and *Milad Jafari* take a broader approach, and propose to reframe the concept of transboundary water security (TWS) toward a more comprehensive and integrated perspective to providing security. They evaluate their framework through the examination of Afghanistan's transboundary river basins. *Ali Oguz Dirioz* keeps the focus on reconsidering widespread understandings of security, dealing with the concept of securitization through the case study of the unsuccessful Water for Peace (WfP) initiative between Türkiye and Israel in the early 2000s.

The third section investigates into the perspectives at the intersections of theory and practice related to the most practical field within the scope of the volume: water diplomacy. *Ayşegül Kibaroğlu* discusses the transformation of the frameworks of water diplomacy focusing on the main actors and institutions involved. She evaluates the changes related to the role of the state and non-state actors, and the emerging cooperative institutional frameworks through the analysis of the Euphrates-Tigris Basin as a case study. *Kinga Szálkai* and *Mary Durfee* apply a more critical approach, discussing the potentials and the shortcomings of multi-stakeholder water diplomacy in addressing issues related to Indigenous rights and sovereignty in the case of salmon and the Columbia River Treaty between the US and Canada. The last chapter in this part, written by *Najibullah Loodin*, *Gabriel Eckstein*, *Vijay P. Singh* and *Rosario Sanchez*, deals with the role of data sharing in transboundary water relations. The authors explore the challenges that prevent riparians from sharing data in an effective and co-operative manner, drawing on the theory of planned behavior (TPB), starting out from the example of the Helmand River Basin to discuss their claims.

The fourth section introduces new perspectives to provide specific entry points for understanding and managing transboundary water conflict and cooperation. *Fatine Ezbakhe* introduces the potential of discourse analysis to unpack how actors define the problems associated with water resources, the solutions considered, and the power relations underlying them, identifying gaps that may serve as future directions for strengthening the discursive approach in transboundary water studies. She illustrates her claims with examples from the Nile, the Jordan, and the Mekong basins. *Péter Marton* navigates rather uncharted waters of research with linking transboundary rivers to public health challenges based on the premise that a One Water Basin/

One Health perspective is necessary for a comprehensive understanding of the nexus between the two areas. His case study explores the related transboundary impact in the Danube River Basin. *Máté Szalai* deals with a topic which is rarely discussed within transboundary water politics, although the logic and the consequences of the presence of shared waters in several cases of small island developing states (SIDS) in seas and oceans do not differ significantly from that of transboundary lake and river issues. This chapter analyses the incentives and limits for cooperation over water between SIDS, highlighting the problems of over-institutionalization, lacking implementation and coordination, and the role of political and economic competition.

Acknowledgements First and foremost, we would like to express our gratitude to the authors of the book. They provided invaluable expertise and perspectives in their chapters, and answered our editorial requests patiently even after numerous rounds of review. We are happy that we share the enthusiasm for transboundary waters and IR theories, and we are indeed grateful for having the pleasure of working together on this project in a friendly and supportive atmosphere.

We would also wish to thank for the insightful comments and suggestions of our colleagues who contributed to the volume with their advice: Fayezullah Azizi, Pál Dunay, Martin Kreuletz, Péter Marton, Tamás Peragovics, Péter Rada, Erzsébet N. Rózsa, András Szalai, Teoman Ertugrul Tulun, and Hayriye Esbah Tuncay. Their expertise and useful critiques improved the book in numerous ways and saved us from many errors.

We are deeply indebted to Viktor Friedmann, our colleague at the Budapest Metropolitan University for his valuable contributions to our work. His extensive personal and professional guidance throughout the whole editorial process is greatly appreciated. We would also like to express our special thanks to Dóra Szabó-Thalmeiner who helped the editorial process with her unique perspective and constructive suggestions.

We would also like to acknowledge the help of our proofreader, Réka Futász. Not only did she provide professional assistance with style and language, her active comments also contributed to the consistency of our work and allowed us to go the extra mile.

This work would not have been possible without the encouragement and the constant support of Erzsébet N. Rózsa and Mary Durfee, and the inspiration provided by the organizers of BPO'19 and the TWIR'21 Team.

References

Conca K, Dabelko G (2002) Environmental peacemaking. Woodrow Wilson Center Press with Johns Hopkins University Press, Baltimore

du Plessis A (2000) Charting the course of the water discourse through the fog of International Relations theory. In: Solomon H, Turton A (eds) Water wars: enduring myth or impending reality. ACCORD/Green Cross International and the African Water Issues Research Unit, Durham/Pretoria, pp 9–34

Elhance AP (1999) Hydropolitics in the Third World. Conflict and cooperation in international river basins. United States Institute of Peace, Washington

Furlong K (2006) Hidden theories, troubled waters. International Relations, the 'territorial trap', and the South African Development Community's transboundary waters. Polit Geogr 25(4):438–458

Gleick PH (1993) Water and conflict. Fresh water resources and international security. Int Secur 18(1):79–112

Gleick PH (2010) Climate change, exponential curves, water resources, and unprecedented threats to humanity. Clim Change 100:125-129

Haftendorn H (2000) Water and international conflict. Third World Q 21(1):51–68
Homer-Dixon TF (1994) Environmental scarcities and violent conflict. Evidence from cases. Int Secur 19(1):5–40
Homer-Dixon TF (1999) Environment, scarcity and violence. Princeton University Press, Princeton
Kipping M (2009) Can 'Integrated Water Resources Management' silence Malthusian concerns? The case of Central Asia. In: Brauch HG et al (eds) Facing global environmental change: environmental, human, energy, food, health and water security concepts. Springer, Berlin, pp 711–723
LeMarquand DG (1977) International rivers: the politics of cooperation. University of British Columbia, Vancouver
Malthus T (1798/1998) An essay on the principle of population. Electronic Scholarly Publishing Project. http://www.esp.org/books/malthus/population/malthus.pdf. Accessed 22 January 2023
Mollinga PP (2001) Water and politics: levels, rational choice and South Indian canal irrigation. Futures 33(8):733–752
Naff T (1994) Conflict and water use in the Middle East. In: Rogers P, Lyndon P (eds) Water in the Arab World: perspectives and prognoses. Harvard University Division of Applied Sciences, Cambridge, pp 253–284
UN Water (2008) Transboundary waters: sharing benefits, sharing responsibilities. https://www.unwater.org/sites/default/files/app/uploads/2017/05/UNW_TRANSBOUNDARY.pdf. Accessed 22 January 2023
Warner JF, Zeitoun M (2008) International Relations theory and water do mix: a response to Furlong's troubled waters, hydrohegemony and international water relations. Polit Geogr 27(7):802–810
Wolf AT (ed) (2013) Transboundary Freshwater Dispute Database. Oregon State University. https://transboundarywaters.science.oregonstate.edu. Accessed 22 January 2023
Wolf AT et al (2003) International waters. Identifying basins at risk. Water Policy 5:29–60

Part I
Revisiting 'Grand Theories'

Chapter 2
Theories of International Relations and the Hydropolitical Cycle: The Hydro-Trap and the Anarchic Nature of Water Conflict

Sameh W. H. Al-Muqdadi

Abstract Transboundary water challenges have prompted by many scholars to investigate the relationship between water conflict/cooperation and International Relations theories. To connect theory with practice, this chapter explores the commonalities between the hydropolitical cycle (HPC) concept and some of the key theories of International Relations based on the case study of the Euphrates and Tiger river basins. The HPC framework is used to outline the possible hydropolitical consequences of water impoundment and connect them to various theories of International Relations (i.e. realism, structural realism, constructivism, and global governance). The results address the connection of these theories within the HPC components where different conflict stages and expected implications of controlling water have been embedded. Building massive dams may lead to inevitable conflicts between riparian countries and influence by regional or international anarchy. The results also show that differences in perception may lead to irrational actions at the state level, thus a global institutional framework is required to cope with the risks of water control. This work recommends a hands-on transboundary policy approach to managing the threats of utilizing massive dams in water impoundment, by adopting the Dam Index concept. A set of criteria are integrated within the Dam Index to rationalize the purpose and operation of dams, alleviate the distrust between riparian countries, and encourage cooperation to empower peacebuilding.

Keywords Dam Index · Hydropolitical cycle · Water conflict · International Relations · Transboundary waters

---

S. W. H. Al-Muqdadi (✉)
Green Charter, Malsch, Germany
e-mail: sameh@green-charter.de

K. Szálkai and M. Szalai (eds.), *Theorizing Transboundary Waters in International Relations*, Springer Water, https://doi.org/10.1007/978-3-031-43376-4_2

2.1 Introduction

The importance of water as a natural resource is recognized at the political level and linked to the socioeconomic and environmental consequences of hydropolitics. Transboundary water conflict/cooperation is reviewed by different international scholars and connected to various classical and critical theories of International Relations (IR) around the world. The present study demonstrates the principle of key theories of IR through the hydropolitical cycle (HPC) concept, which addresses the consequences of water control/conflict between riparian countries. The work includes a case study of water impoundment by a dam complex in the Euphrates-Tigris basin. The research connects theory with practice by associating different traditions of IR theories with various parts of the HPC and presenting the impact of dams in escalating water conflicts between the riparian countries of the Euphrates-Tigris basin.

The HPC framework (Al-Muqdadi 2019) shows potential interconnections among the disagreements/disputes over the shared basins of riparian countries, the consequences of controlling water by massive dams, and the possible roles of IR theories to shape the relation between the riparian countries. HPC includes five major phases: national security, food security, environmental issues, social concerns, and economic challenges. The five phases are built on a synthesis of different IR theories, where the theories are embedded within the HPC phases to emphasize the practical connections between water conflict and various IR theories.

The study develops the HPC concept by elaborating on the possible stages of conflict, integrating key theories of IR, and illustrating the political weight and perception that commands hydropolitical events. The development of HPC turns the complex hydropolitical system into a more dynamic one, leads to a better understanding of transboundary water challenges, and reveals the IR theories' interventions on the riparian countries perception and actions at the different proposed phases of escalating water conflict. The ultimate objective of the current chapter is to underline the inherent implications of massive dam projects on regional and international security. As watercourses create challenges that cross political borders, this study suggests a transitional governance policy to mitigate the risk of water conflict, adopting the concept of a Dam Index (DI), which may help control the construction of massive dams and make the dams more productive for riparian countries. The DI also proposes a set of potential areas that need to be considered as criteria to rationalize massive dam operations. The chapter begins with an extensive literature review of the last two decades of International Relations and transboundary water research, followed by a case study of the Euphrates and Tigris River basins. The concept of the hydropolitical cycle is also demonstrated to illustrate the consequences of water impoundment. The results will practice the impact of theories of IR along with the transboundary water conflict potential phases within the HPC.

2.2 Research Context

The present research builds on two academic research programs: IR theory and transboundary water politics, and the literature on the case study of the Euphrates and Tigris Rivers basins.

2.2.1 *Two Decades of International Relations and Transboundary Water Research*

While the study of transboundary water politics in IR has only become popular since the 1990s, traditional ideas of the International Relations literature are also used in the present study, exploring elements of classical realism and the Thucydides trap (Forde 2012) to disclose the root causes of the inevitable conflict over water resources. Thucydides, the father of classical realism, believed that war was inevitable and driven by fear. He lent his name to the notion of the Thucydides trap (Allison 2017), the idea that a rise in the power of one country will be perceived as a threat by a neighboring country, which will then attempt to increase its power, inevitably leading to war.

The notion of anarchy and how it affects the complex system of hydropolitics will match with the core principle of structural realism (Mearsheimer 2007). The absence of a central authority leads to a situation where no policies are enforced, and states become insecure and obsessive about survival. Morgenthau identifies conflict as rooted in anarchy (Speer 1968).

The role of epistemology is important to highlight the differences in identities and norms barriers that may expand the communication gap between riparian countries, and the theory of constructivism (Wendt 1995) can be used to clarify this challenge. According to Alexander Wendt, the founder of constructivism, states might act as irrational individuals if they persist with a sovereignty regime despite the costs and refuse to recognize others outside their borders (Wendt 2005), with different identities and norms shaping perceptions (Morris et al. 2015).

Representatives of Marxism, and other scholars who address the economic dimension as crucial in shaping the behavior of the state (Worsley 2013) are used in the current research to reflect the social and economic pressure caused by hydropolitical events. Since inter-state transboundary water is a global challenge that crosses political borders, the global governance concept has underlined the way forward of how the states should deal with the transboundary challenges (Rosenau 2006). Here, Rosenau's global governance concept helps us understand the collective management of transnational or international problems that cross political borders, such as climate change, terrorism, pandemics, water resource scarcity, and the sharing of catchments (Payus 2020).

Global research efforts during the last couple of decades in terms of water and International Relations show outstanding insights from different scholars in different regions. Fox (1993) provides one of the earliest studies to link water scarcity with international relations. According to the author, environmental topics were relatively neglected at the time, and the theoretical framework presented is coupled with a case study on the Euphrates and Jordan Rivers, seeking to prove a link between environmental degradation and interstate conflicts. The results show an increase in political tension and antagonism between riparian countries. A few years later, S. Dinar (2000) analyzes water-related conflict and cooperation in the context of international relations and presents several case studies, including one on the Euphrates-Tigris basins (ETB) and the Nile basin. Dinar's work illustrates the interventions and interactions of variables among the negotiating parties and concludes that water can be a key variable in international relations, and it is fundamental to building peace or unleashing conflict. On the other hand, Dyer (2001) examines the concept of environmental security and its relevance to international relations. The author concludes that the construction of any long-term international perspective on interstate global politics could be quite a difficult mission. Stahl (2005) investigates the impact of hydroclimatic, socioeconomic, and political drivers on the pattern of water-related international relations issues. The study analyzes global data covering the period 1950–2000 and clusters relevant political events in tree models. Stahl's results show that the cooperative pattern was primarily dominant, while the highest proportion of water-related conflicts appeared in the Middle East and Asia.

In 2006, Furlong (2006) carried out a case study of shared water resources in the Southern African Development Community, which includes 16 riparian countries. The author hypothesizes that delving into the relationship between watercourses and International Relations theories might neglect some challenges and risks, obfuscating many more. Warner and Zeitoun (2008) present an argument on the general influence of International Relations in the context of water conflict or water cooperation. Their study focuses on neo-realist and neo-liberal approaches to water policies, particularly in the Middle East and South Africa. The authors highlight the lack of academic efforts to discuss International Relations in hydropolitics in depth, especially when addressing essential challenges in terms of power distribution and hegemony in water politics. Wolf (2009) discusses the water cooperation between riparian states from a different angle, asking the crucial question why countries that share a basin cooperate on water, even when they will not cooperate over other issues. Wolf believes that water is a subject of conflict and that theories have failed to explain cases with water-related cooperation.

Asking the key question "is water politics?", Warner and Wegerich (2010) present an extensive literature review of studies with a global scope on water as a political tool. The study highlights the lack of a solid definition for the term 'water policy', since only a few studies have focused on the binding relationship between water and international relations. A 2015 study by Pereira (2015) discusses the characteristics of the Westphalian and post-Westphalian international relations systems, arguing for the importance of natural resources in the context of international relations and the inherent effects of globalization. The author encourages scholars to spread the positive impact of developing an integrated global discipline to mitigate potential global security risks. Menga (2016), on the other hand, applies an analytical framework to the interaction of domestic politics and transboundary water politics and claims that the interactions between domestic and foreign policy are rarely investigated, and that collective research in this area would create a better understanding of water politics. Menga's framework covers the literature on nationalism and the impact of large dams, e.g. the Grand Dam in Ethiopia and the Rogun Dam in Tajikistan. Shroder and Ahmadzai (2016) describe the political approach, cultural aspects, and economic implications of the transboundary water resources in the Kabul River and the Indus watershed. Their study provides a local water management case study and includes comprehensive negotiations rounds between the riparian countries in Southwest Asia.

Warner et al. (2017) introduce a hydro-hegemony framework using the International Relations theories of realism and neo-institutionalism. Their research indicates that transboundary water relations are non-linear and provides a conceptual model with which to illustrate the negative impacts of hydro-hegemony. Schmeier and Shubber (2018) discuss creating a framework of International River Basin Organizations (RBOs) to stabilize the cooperation between riparian countries for the long term. More recently, Loodin and Wolf (2021) highlight the principle of water management in a dimension represented by religion. They use Islamic Sharia law by implementing three criteria: equity, responsibility, and sustainability. They investigate the Helmand River treaty with Iran to determine if Islamic principles are implemented to manage water resources and the construction of dams under the Taliban regime.

2.2.2 *A Case Study of the Euphrates and Tigris River Basin (ETB)*

As a cradle of civilization, the Euphrates and Tigris River basin has a long history of water conflict. As the ultimate downstream country of the ETB, about 98% of Iraq's surface water is supplied from the Tigris and Euphrates Rivers, with over 90% of the water contribution supplied from the riparian countries of Turkey, Syria, and Iran. About 70% of this water contribution is from Turkey as an upstream country for the ETB (Iraq 2013). Early transitional water disagreement happened right after World War I, when Ottoman empire fall and new political borders were set to shape the current riparian countries Iraq, Turkey, Syria and Iran (Al-Muqdadi et al. 2016). However, the region has failed to reach a strategic agreement to manage the transboundary water for both Euphrates and Tigris basins for a century. Obtaining a mutual agreement is still challenging, as most treaties have been concluded based on acute political requests (Al-Muqdadi 2019). There are several reasons for the regional struggle to achieve a water agreement on the ETB. First, the international watercourses laws of 1997 (UN convention of non-navigational use of watercourses) give the same weight to the past as to the present (Priscoli and Wolf 2009). Second, there is a lack of agreement whether to define the Euphrates and Tigris Rivers as international rivers (as Iraq and Syria claim) or as transboundary watercourses (as Turkey claims) (Kibaroglu and Scheumann 2011). Third, it is debated whether the rivers should be treated as one basin (as Turkey and Syria propose) or as two separate basins (as Iraq proposes) (Harris and Alatout 2010). Furthermore, there is a disagreement on the proper water allocation formulas (Kibaroğlu et al. 2007), as well as different perspectives and norms, where Turkey emphasizes sovereignty, Iraq highlights the Mesopotamian heritage, and Syria expresses food security concerns (Al-Muqdadi 2019). Finally, the fragile political environment of the region and the lack of acceptable mediation also lead to miscommunication between the riparian countries, who rarely meet with each other (Eroglu 2018).

The Euphrates-Tigris basins have passed through two remarkable political events in the last two decades, particularly in the case of the downstream countries, as the invasion of Iraq in 2003 and the Syrian uprising in 2011 have caused severe political instability for both. The upstream countries (Turkey and Iran) have gained the advantage of this asymmetric power to build massive dams, which has resulted in severe water supply cut-off for Iraq and Syria. For example, the Southeastern Anatolia Project (GAP) was established to build 22 dams, 19 hydroelectric plants, and extensive irrigation systems on the Tigris and Euphrates Rivers. The estimated water supply reductions in Syria and Iraq are 40% and 80%, respectively (Zarei 2020). Given the observed impact of climate change in the region, which is causing water stress, the possibility of a dispute over water resources will increase dramatically in the coming years.

2.3 The Hydropolitical Cycle

The hydropolitical cycle is a complex nexus of events that shape the relationship between upstream and downstream countries. Where more actors are involved, a more complicated system is expected. The loop starts with the upstream country building dams as a national development initiative. Coupled with the absence of cooperation and communication, dams may have severe consequences for the downstream country. The possible escalation of water conflicts leads to five phases: (1) the national security phase includes controlling water by massive dams and leads to a distrust between the riparian countries, with sovereignty often used by the upstream country as justification. (2) In the food security phase, water shortages lead to irrigation challenges and shrinkage in the agriculture sector. (3) This leads to environmental issues, where the downstream country suffers from desertification, ecosystem degradation, and health concerns. (4) Social concerns are represented by a displacement wave of farmers heading to the main cities or immigrants crossing state borders. (5) The final phase includes economic challenges, where main cities become crowded, the infrastructure is overloaded, and an economic crisis is expected. Driven by political decisions, the fifth phase may later lead to a conflict between the riparian countries.

The framework also reflects riparian countries' political interception and perceptions through different conflict stages to express the potential level of the water dispute. Pondy's description of conflict stages (Pondy 2011) explains the nature and status of possible disagreement and conflict between the riparian actors within different phases of the HPC framework. Figure 2.1 represents the Pondy's six conflict stages, which include latent, tension, perceived, felt, manifest and aftermath that would reflect the level of water conflicts between the riparian countries within the HPC.

2.4 Results and Discussion

Practicing of the theories of IR impact on the transboundary water conflict is reflecting vigorous links with the HPC phases and components as following.

2.4.1 The National Security Phase and the Hydro-Trap

Water discord over the sharing of catchments may take different forms through different actions on the part of the upstream countries, such as water impoundment by building dams, unilateral action that leads to water diversion from a natural water body, and upstream pollution from industry or mining operations. The current research focuses on the type of water control that uses dams.

The generic model of the HPC fits any shared-basin conflict where the upstream party controls water through massive dam projects. Such action can potentially transmit signals to the downstream party that may result in distrust, particularly when the upstream country invokes the notion of its sovereignty in its national development plans. The conflict stage at this level may become latent where conflict is still hidden or unforeseen; however, this phase might represent national security. The expectation of water shortages can influence the agricultural sector and lead to a phase of food security, where tension reflects the emotional strains effected by distrust and fear. Controlling water by utilizing massive dams could be a modern form of the Thucydides trap (Forde 2012), where upstream countries will leverage their power by controlling water resources. *Managing and Transforming Water Conflicts* by Priscoli and Wolf (2009) describes the advantages of geopolitical setting and a riparian position, where controlling water by dams gives upstream nations positions of greater power, and creating a flash point for regional conflict. For example, despite the asymmetric power between the upstream countries Turkey on the ETB, and Ethiopia on the Nile basin, both countries gain substantial power by controlling watercourses with dams. Controlling water upstream could feed hydro-distrust, where the perceived fear and uncertainty by the downstream countries will increase dramatically. In some cases, the hydro-trap implications go even further, leading to a proxy war, for example, Syria has made use of the rebel troops, the Kurdistan Workers' Party

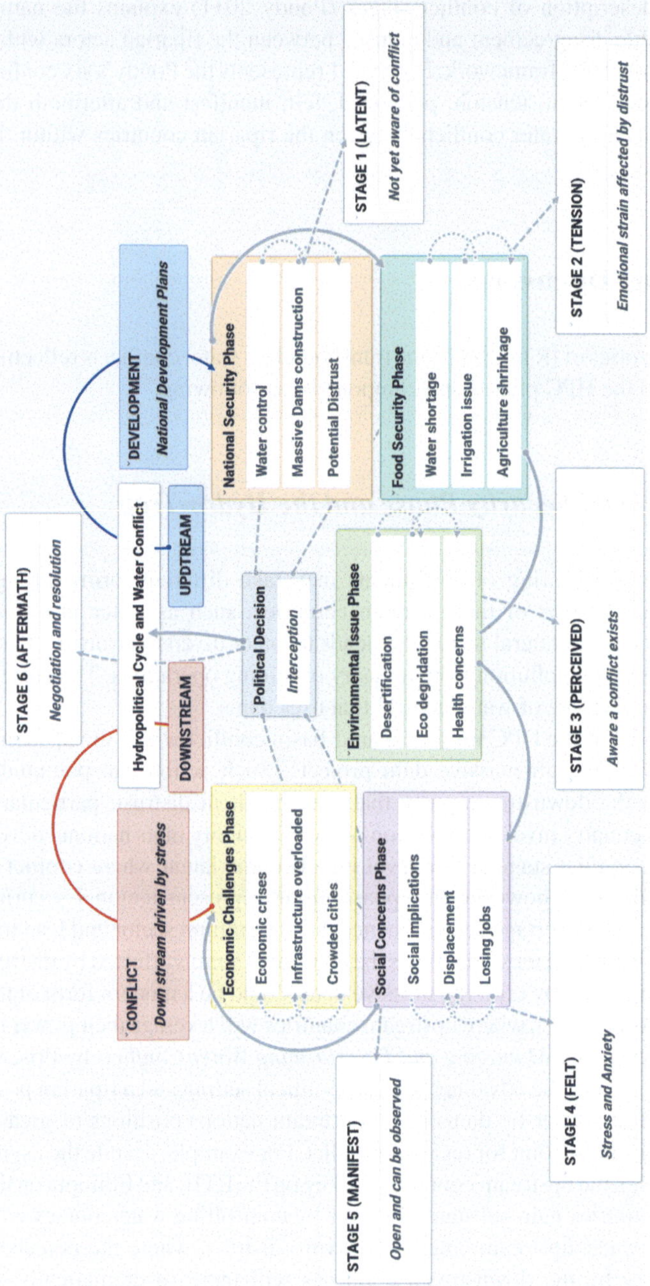

Fig. 2.1 The hydropolitical cycle and potential stages of conflict, following Al-Muqdadi (2019)[1]

[1] The open-source Plectica software was used to visualize the conceptual foundation of the HPC phases (Plectica LLC 2020).

(PKK), to force Turkey to negotiate water allocation from the Euphrates River for decades (Daoudy 2009).

Nevertheless, the assumed hydro-trap might not lead to war, but it could work as a trigger for inter-state water competition and create an unhealthy environment that prevents possible water-share treaties, which can collectively lead to a water conflict. For example, the water disputes in the ETB region emerged after a series of dams were constructed by Turkey on the Euphrates and Tigris Rivers.

2.4.2 The Food Security and the Environmental Challenges Phase

The Southeastern Anatolia Project (GAP) has speeded up the escalation loop of the water confrontation between riparian countries (Pool and Grover 2006). GAP has always been a source of tension between Turkey and Syria, as well as a lesser tension between Turkey and Iraq, due to their strategic security cooperation against the PKK (Dohrmann and Hatem 2014). The massive dams will cause profound uncertainty in the political environment, and a significant shortage of water supply will lead to food security concerns and environmental challenges. A range of implications are expected: the shrinkage of the agriculture sector, increased desertification, ecological degradation, and an increasing number of health issues. At this point, the water conflict stage will be perceived where at least the downstream state is aware that conflict exists.

Coupled with the influence of climate change and the rapid growth rate of the region, the estimated degradation of irrigated land in Iraq and Syria will be 71% and 17%, respectively (Dregne and Chou 1992). The areas of salt-affected soil are estimated to be 61% in Iraq and 40% in Syria by 2050 (Haktanir et al. 2004). The water shortages have also affected the downstream countries' operation of their hydropower plants (TILMANT 2007), which puts downstream countries at risk of losing the opportunity to use clean energy and achieve long-term sustainable development goals.

2.4.3 The Social Concern Phase and the Anarchy of the Complex System

Having agreements does not necessarily mean that the riparian states are already cooperating on water. Nevertheless, suppose the riparian countries reach no agreement. In that case, the implications are likely to escalate to the phase of social concern, where people begin to lose jobs due to the degradation of the agriculture sector, and the state may experience waves of displacements that can lead to social problems. At this stage, the riparian countries experience conflict, which provokes stress and anxiety. About 12 million people in Syria and Iraq are currently threatened by displacement due to drought and water crises (Devi 2021). Crop failure and the loss of cattle in Iraq are estimated at 30% and 37%, respectively, and young people are losing jobs or leaving home to look for economic opportunities elsewhere (NRC 2021). Although the social implications could already be there, the question is rather how severe they are and how much the water conflict would contribute to them, especially since the region has been suffering from political instability.

The HPC reflects the fact that there is a lack of a solid structure for decision-making and the timeframe for reaching each phase. However, controlling water by massive dams could be one of the reasons that reflect the consequences of that anarchy at a regional or international level, Morgenthau highlights (Morgenthau 1945). Anarchy leads to poor water management at a state level, which would add further challenges to the existing transboundary water issues. However, an in-depth investigation is required to determine whether controlling water contributes to anarchy and, if so, to what extent and at what cost.

The ETB is a region characterized by dynamically rich geopolitics and critical events, a regional instability that could contribute to international anarchy. During the last two decades, Iraq has passed through a radical shift from deep centralization to partial decentralization, with insufficient time to prepare for such a fundamental shift. The Iraqi experience has resulted in political impediments and policy confusion, with a lack of rules to define authority and procedures (Fleet 2019). This poorly executed to decentralize the political system is reflected in different fields at the state level, one of them being water governance. The Iraqi constitution, and the contradiction between Articles 110 and 114 indicate a struggle between centralizing and decentralizing the political system. On the one hand, Article 110 establishes a federal control of water resources; on the other hand, Article 114 gives the right to manage and control water to the regional and local levels, leading to a conflict of interest between the local and the federal government (Iraqi Government 2005). Another example of regional anarchy is when Turkey's response to the Arab Springs adopted the zero-problem philosophy (ZPP) to solve its challenges with their neighboring countries and to establish a new era with neighboring countries in terms of foreign policy. However, the ZPP doctrine does not work in Syria due to the regional political challenges and the potential risk of instability, especially with the presence of the PKK, which always represents an imminent threat to southern Turkey (Zalewski 2013; Chiriatti and Donelli 2015). Controlling water by the GAP project could be a puzzle piece in the failure of the ZPP.

2.4.4 The Economic Challenges Phase and the Obsessions of Surviving at State Level

It is expected that these snowballing consequences of the social concerns phase may result in economic challenges. Cities may become more crowded due to the waves of displacement, the infrastructure may become overloaded, and a financial crisis may engulf the state. Between December 2007 and June 2009, 25,578 people were displaced due to drought in Iraq (Lück 2014). At this point, the water conflict stage between the riparian countries can be manifested, in which conflict becomes open and observable.

Scholars of Marxism and other scholars of the political economy indicate how the economy trumps other concerns and how it can be the driving force of conflict (Taylor 2008). Following a similar scenario to Turkey's water control through GAP, Iran has established the Tropical Water Project (TWP), which includes 13 dams with a capacity of 1.7 billion cubic meters and 150 km of tunnels to divert the water to the rural area of southern Iran for irrigation and economic development purposes (Sarhat 2022). The TWP has caused a decline of 70% in the water supply of the Lower Zab River in Iraq (Rudaw 2018). However, there is an Iraqi public concern that Iran is thus boosting its food exports by cutting the Iraqi water supplies, since Iraq has been the destination for 35% of Iran's agricultural exports (Deutsche Welle 2021). The drought has led many Iraqi activists to denounce the Iranian actions and express disappointment towards the Iraqi government's lack of action (Abdulrahman 2017).

Downstream states may be under significant pressure to ensure economic survival and deal with possible frustration among the public, which might lead to irrational actions, as Wendt describes in his theory of constructivism (Wendt 1995), where the differences in identities and norms might fuel the conflict of the nation. This explains why the global struggle over water resources varies across regions and could be one of the reasons why the international community witnesses different behaviors in similar situations. The building of massive dams in shared basins can be uncontroversial in region X and highly controversial in region Y due to multilateral gaps, such as an insufficient understanding of cultural sensitivity, a lack of communication, differences in the level of cooperation, and differences in the tenets of International Relations approaches adopted in each region. Consequently, considering all the gaps mentioned above between upstream and downstream states, and if the downstream state adopts the realist theory of International Relations, a toxic situation can result, where a water conflict is more likely to occur.

Thanks to constructivism, this mechanism can be seen in the dynamic system of the HPC, where irrationality is highly expected after the economic challenges phase. Iraq and Syria were on the brink of a possible water war in 1975 due to the water flow reduction in the Euphrates River when the Keban Dam was filled up in Turkey (Kolars 1994). In 1992, the Pentagon reported a water war between Syria and Turkey as one of the first contingencies of the water dispute of the Euphrates River (Dolatyar and Gray 2000).

2.4.5 The Post-conflict Phase and the Concept of Global Governance

There is a time when water conflicts end, and the aftermath stage involves seeking mutual interests for the common good, representing the capacity for responsible leadership. At this stage, Rosenau and the theory of global governance (Rosenau 2006) help extend a better understanding to the level of the decision maker and encourage them to cooperate over transnational challenges that cross political borders, such as transboundary water issues. The aftermath stage of water conflict presented on ETB by few strategic agreements was conducted with help from regional or international mediation. The Adana agreement in 1998 was a rare strategic agreement between Turkey and Syria, which helped both countries cooperate in different fields, and the water supply to Syria doubled to 982 m^3/sec (Carkoglu and Eder 2001), while the Iraq–Iran political settlement over the Shatt-Al-Arab River by way of the March 1975 Algiers Agreement attempted to solve the boundary dispute between the two neighboring countries (Biger 1989). However, agreements could be terminated at any time by political force due to the fragile nature of the region, its complex political system, and the lack of a comprehensive institutional framework.

International Relations theories assist in understanding how international systems work and explain the reasons beyond peace and conflict for a state's behaviors, whether they involve state empowerment, institutional enablement, or economic development. This International Relations philosophy for classical theories, which relies on geopolitical facts and the limited political borders of the state, contradicts the basic philosophy of water as an element. Watercourses neither appreciate political boundaries nor care about the behavior of states or institutions. Watercourses are not driven by economic aspirations, they follow two simple sets of rules: gravity and pressure differentials. Notably, the notions of water philosophy and International Relations are brought together by global governance theory. Problems can only be confronted efficiently when global governance allows for the discussion of cross-border challenges (such as transboundary water on a broader scale) and a capacity for proper awareness of environmental challenges at the global level.

2.4.6 The Concerns of Water Conflict Legacy and the Way Forward

Operating massive dams on watercourses without proper consultation and cooperation with downstream states or the consideration of further consequences is the epitome of power acquisition. It shifts the control of water from communities to governments and colonizes both rivers and people (Shiva 2016). This approach to large-scale water projects can unleash conflict. Many scholars highlight the disadvantages of dams, particularly those built on a large scale. Dams are believed to negatively influence health and food security by increasing communicable diseases and social disruption (Lerer and Scudder 1999), and massive dam structures are most likely to erode and may be affected by climate change, which puts downstream areas at risk of flood if they fail (Hughes and Hunt 2012).

Beside the negative ecological impacts of the dams (Kuriqi et al. 2020), *The Greater Common Good* (Roy 1999) emphasizes that massive dams are to national development what nuclear bombs are to military arsenals: they are both weapons of mass destruction (WMD). Unlike WMD, however, the negative impact of massive dams is slow to manifest, although the legacy of collateral damages could remain for the long term. In this context, constructing massive dams without restrictions or at least firmer international policies could prompt threats and global concerns similar to the spread of WMD, which should be monitored and controlled. Therefore, there is an acute need to rationalize and control large-scale dams in shared catchments to ensure that the fundamental aim of managing the water resources will not be shifted into a threat to downstream countries. The international community and the United Nations Office for Disarmament Affairs are making substantial efforts to support the disarmament of states that possess WMDs, and outstanding strategies have been offered to enhance the role of regional organizations in preventing WMD proliferation (Shirazyan 2018). There is a similar pressing need to control massive dams and their impacts by having more international legislation and developing policies that cross political borders. This includes governing international watercourses to mitigate the risk of misusing dams for political agendas and to alleviate the collateral damage dams create.

The present study proposes the concept of a Dam Index (DI), recommending it for adoption by international institutions such as the United Nations. The DI is a set of requirements that should be fulfilled whenever a large-scale water project in a shared catchment is planned for construction. Five major steps are needed: (1) preparation, where the developer country provides sufficient evidence of consultation and cooperation with the downstream state and undertakes an environmental assessment; (2) permission, where global due diligence is performed to fulfil criteria in terms of the size, purpose, and expected impact of the dam; (3) regulation, where the stakeholders and policies related to the water project are mapped in the context of the conventions set out in the Law of the Non-Navigational Uses of International Watercourses (McCaffrey 2008); (4) monitoring, where an independent global organization (e.g. United Nations–Water) remotely observes the operation to guarantee transparency, good practice, and responsible use; and 5. rating, where a scoring system is used to regularly evaluate concerns about the project commitments over the years. The notion of sovereignty should not be seen as superseding regional or international security issues. The practical implementation of the DI would have several advantages, including (a) controlling the risk of water conflict between riparian countries by having a clear policy guideline to stop the frantic race to build massive dams worldwide; (b) providing a solid structure and reporting system to mitigate the distrust between the riparian countries and encourage cooperation to empower peacebuilding; (c) rationalizing the purpose of dams and their operation; (d) including a monitoring system to assist in achieving a bigger goal beyond water security, having regional cooperation between riparian countries to confront the common external challenge of climate change. This approach would also serve the global strategy of climate change mitigation and adaptation in the long run.

2.5 Conclusion

The components of the five phases of the hydropolitical cycle (national security, food security, environmental issues, social concerns, and economic stability) are built on a synthesis of various theories of International Relations. Classical realism is relevant to explaining and understanding the early phases, while structural realism, Marxism, constructivism, and liberalism/global governance help illustrate the other phases. From the realist perspective of International Relations, massive water projects such as large-scale dams in shared catchments reflect national security issues, and they increase distrust between the riparian countries and feed the root cause of water conflicts. Continuing to ignore water crises without going into sustainable solutions and strategic agreements leads to an increased possibility of conflict. Controlling water though dams is a trigger for raising power and could be identified as a hydro-trap, since water impoundment speeds up the escalation loop of water confrontation between the riparian countries involved, resulting in evident challenges at different levels of environmental security, as well as socio-economic concerns.

The complex systems present in the HPC reflect the anarchy that Morgenthau highlights in structural realism theory. This anarchy could contribute to water conflict and increase the risk of instability and insecurity. Given the presented challenges of controlling water, differences in norms and identities may play a role in making a state act irrationally. Wendt's critical theory of constructivism supports this conclusion, where states act as irrational individuals if they persist in the sovereignty regime despite the costs and refuse to recognize others outside their borders. However, the principles of watercourse philosophy and theories of International Relations can come together through global governance theory, which helps understand the collective management of common transnational or global problems that cross political borders, with the capacity of rational leadership for building peace reflected in the post-conflict stage. Therefore, more transnational governance and a modern monitoring system should regulate massive dams.

The Dam Index is proposed as a global policy guideline, as a practical approach to stop the frantic race to build massive dams worldwide to control water resources in shared catchments. The DI would help promote a clear structure to rationalize the purpose of having dams, alleviate the risk of distrust between transboundary water nations, encourage cooperation for transitional global challenges such as the impact of water and climate change, and boost peacebuilding strategies.

References

Abdulrahman SA (2017) Failure to securitise environmental problems in Iraq: the case of the Alwand and Sirwan Rivers. Int J Environ Stud 74:540–557. https://doi.org/10.1080/00207233.2017.1330493

Allison G (2017) The Thucydides trap. Foreign Policy 9(6)

Al-Muqdadi SWH (2019) Developing strategy for water conflict management and transformation at Euphrates-Tigris basin. Water 11(10):2037. https://doi.org/10.3390/w11102037

Al-Muqdadi SW, Omer MF, Abo R, Naghshineh A (2016) Dispute over water resource management—Iraq and Turkey. J Environ Prot 7(8):1096–1103

Biger G (1989) The Shatt-Al-Arab river boundary: a note. Middle East Stud 25(2):249–252. https://doi.org/10.1080/00263208908700779

Carkoglu A, Eder M (2001) Domestic concerns and the water conflict over the Euphrates-Tigris river basin. Middle East Stud 37(1):41–71. https://doi.org/10.1080/714004364

Chiriatti A, Donelli F (2015) Turkish 'zero problems' between failure and success. Res Turk 4(3):108–131

Daoudy M (2009) Asymmetric power: negotiating water in the Euphrates and Tigris. Int Negot 14(2):361–391. https://doi.org/10.1163/157180609X432860

Deutsche Welle (2021) Iranian dams cut off Iraqi water supplies. https://www.dw.com/en/tensions-rise-as-iranian-dams-cut-off-iraqi-water-supplies/a-58764729. Accessed 31 January 2022.

Devi S (2021) Syria and Iraq facing severe drought. Lancet 398(1395). https://doi.org/10.1016/S0140-6736(21)02238-8

Dinar S (2000) Negotiations and international relations: a framework for hydropolitics. Int Negot 5(2):375–407. https://doi.org/10.1163/15718060020848712

Dohrmann M, Hatem R (2014) The impact of hydro-politics on the relations of Turkey, Iraq, and Syria. Middle East J 68(4):567–583

Dolatyar M, Gray TS (2000) The politics of water scarcity in the Middle East. Env Polit 9(3):65–88. https://doi.org/10.1080/09644010008414538

Dregne HE, Chou N-T (1992) Global desertification dimensions and costs. Degradation & Restoration of Arid Lands, Texas Tech University, Lubboch, pp 73–92

Dyer H (2001) Environmental security and international relations: the case for enclosure. Rev Int Stud 27(3):441–450. https://doi.org/10.1017/S0260210501004417

Eroglu D (2018) Addressing inadequate transboundary water cooperation in the Euphrates and Tigris basin: the guidance of systems thinking

Fleet M (2019) Decentralization and its discontents in Iraq. Middle East Institute. https://www.mei.edu/sites/default/files/2019-09/Decentralization%20and%20its%20Discontents%20in%20Iraq.pdf. Accessed 31 January 2022

Forde S (2012) Thucydides and 'realism' among the classics of International Relations. In: Harloe K, Morley N (eds) Thucydides and the modern world. Cambridge: Cambridge University Press. https://doi.org/10.1017/CBO9781139094214.009

Fox JD (1993) The politics of water scarcity in the Euphrates and Jordan River Basins. MA Thesis, McGill University

Furlong K (2006) Hidden theories, troubled waters: International Relations, the 'territorial trap', and the Southern African Development Community's transboundary waters. Polit Geogr 25(4):438–458. https://doi.org/10.1016/j.polgeo.2005.12.008

Haktanir K, Karaca A, Omar SM (2004) The prospects of the impact of desertification on Turkey, Lebanon, Syria and Iraq. In: Marquina E (ed) Environmental challenges in the Mediterranean 2000–2050. Springer, Dordrecht, pp 139–154

Hughes AK, Hunt D (2012) A guide to the effects of climate change in dams. In: Dams: engineering in a social and environmental context: Proceedings of the 17th conference of the British Dam Society, London, pp 231–246

Iraqi Government (2005) Iraqi constitution. https://www.constituteproject.org/constitution/Iraq_2005.pdf?lang=en. Accessed 31 January 2023

Kolars J (1994) Problems of international river management: the case of the Euphrates. In: Biswas AK (ed) International waters of Middle East from Euphrates-Tigris to Nile. Oxford University Press, Oxford, pp 45–94

Kibaroglu A, Scheumann W (2011) Euphrates-Tigris rivers system: political rapprochement and transboundary water cooperation. In: Turkey's water policy: national frameworks and international cooperation. Berlin, Heidelberg: Springer Berlin Heidelberg, pp 277–299

Kibaroğlu A, Çakmak B, Doğan A (2007) Global water policies and river basin management: reflections on water resources management in Turkey. In: International Congress on River Basin Management, Antalya, Turkey. http://www2.dsi.gov.tr/english/congress2007/chapter_3/76.pdf. Accessed 05 May 2011

Kuriqi A, Pinheiro AN, Sordo-Ward A, Garrote L (2020) Water-energy-ecosystem nexus: balancing competing interests at a run-of-river hydropower plant coupling a hydrologic–ecohydraulic approach. Energy Convers Manage 223(113267):1–19. https://doi.org/10.1016/j.enconman.2020.113267

Lerer LB, Scudder T (1999) Health impacts of large dams. Environ Impact Assess Rev 19(2):113–123

Loodin N, Wolf AT (2021) Will Islamic water management principles be included if the Helmand River Treaty is revisited? Water 14(1):67. https://doi.org/10.3390/w14010067

Lück A (2014) Integrated drought risk management—DRM—national framework for Iraq: an analysis report. https://unesdoc.unesco.org/ark:/48223/pf0000228343. Accessed 31 January 2023

McCaffrey SC (2008) Convention on the Law of the Non-Navigational Uses of International Watercourses. United Nations. https://legal.un.org/avl/pdf/ha/clnuiw/clnuiw_e.pdf. Accessed 31 January 2023

Mearsheimer JJ (2007) Structural realism. In: International Relations theories: discipline and diversity 83:77–94

Menga F (2016) Domestic and international dimensions of transboundary water politics. Water Altern 9(3):704–723

Morgenthau HJ (1945) Nationalities and national minorities

Morris MW, Hong YY, Chiu CY, Liu Z (2015) Normology: integrating insights about social norms to understand cultural dynamics. Organ Behav Hum Decis Process 129:1–13

NRC (2021) Iraq's drought crisis and the damaging effects on communities. https://www.nrc.no/resources/reports/iraqs-drought-crisis-and-the-damaging-effects-on-communities. Accessed 31 January 2023

Pereira JC (2015) Environmental issues and international relations, a new global (dis) order-the role of International Relations in promoting a concerted international system. Rev Bras Política Int 58(1):191–209. https://doi.org/10.1590/0034-7329201500110

Payus C, Ann Huey L, Adnan F, Besse Rimba A, Mohan G, Kumar Chapagain S, Roder G, Gasparatos A, Fukushi K (2020) Impact of extreme drought climate on water security in North Borneo: case study of Sabah. Water 12(4):1135

Pondy LR (2011) Organizational conflict: concepts and models. In: Godwyn M, Gittell JH (eds) Sociology of organizations: structures and relationships. Sage, London, pp 427–439

Pool AM, Grover VI (2006) GAPs in the dialogue of governance: conflicting ideologies of development in Turkey. In: Grover VI (ed) Water—global common and global problems. CRC Press, Boca Raton, pp 373–397

Priscoli JD, Wolf AT (2009) Managing and transforming water conflicts. Cambridge University Press, Cambridge

Rosenau JN (2006) Global governance as disaggregated complexity. In: Contending perspectives on global governance. Routledge, pp 131–153

Roy A (1999) The greater common good. Frontline. https://web.cecs.pdx.edu/~sheard/course/Design&Society/Readings/Narmada/greatercommongood.pdf. Accessed 31 January 2023

Rudaw (2018) Iraq water crisis: Little Zab River shrinks by 70%. https://www.rudaw.net/english/kurdistan/090620182. Accessed 31 January 2023

Sarhat AR (2022) Iranian Tropical Water Project on Sirwan River and its impacts on some environmental aspects in Garmian Region. Journal of Garmian University 9(Special Issue):272–287
Schmeier S, Shubber Z (2018) Anchoring water diplomacy—the legal nature of international river basin organizations. J Hydrol 567:114–120
Shirazyan S (2018) 'Synergies of strengths': a framework to enhance the role of regional organizations in preventing WMD proliferation. Arms Control Today. https://www.armscontrol.org/act/2018-09/features/%E2%80%98synergies-strengths%E2%80%99-framework-enhance-role-regional-organizations-preventing. Accessed 31 January 2023
Shiva V (2016) Water wars: privatization, pollution, and profit. North Atlantic Books, Berkeley
Shroder JF, Ahmadzai SJ (2016) Transboundary water resources in Afghanistan: climate change and land-use implications. Elsevier, Cambridge
Speer JP (1968) Hans Morgenthau and the world state. World Politics 20(2):207–227
Stahl K (2005) Influence of hydroclimatology and socioeconomic conditions on water-related international relations. Water Int 30(3):270–282. https://doi.org/10.1080/02508060508691868
Taylor M (2008) Power, conflict and the production of the global economy. In: Taylor M (ed) Global economy contested: power and conflict across international division of labour. Routledge, London, pp 11–33
Tilmant A (2007) Impacts of the Southeastern Anatolia Project in Turkey on the performance of the Tabqa dam and hydropower plant in Syria. In: Management, Proceedings of Symposium HS3006 at IUGG2007, Perugia, 2007, pp 92–99
Warner JF, Wegerich K (2010) Is water politics? Towards international water relations. In: Warner JF, Wegerich K (eds) The politics of water. A survey. Routledge, London, pp 3–18
Warner JF, Zeitoun M (2008) International Relations theory and water do mix: a response to Furlong's troubled waters, hydro-hegemony and international water relations. Polit Geogr 27(7):802–810. https://doi.org/10.1016/j.polgeo.2008.08.006
Warner J, Mirumachi N, Farnum RL et al (2017) Transboundary 'hydro-hegemony': 10 years later. Wiley Interdiscip Rev Water 4(6):1–13. https://doi.org/10.1002/wat2.1242
Wendt A (1995) Constructing international politics. Int Secur 20(1):71–81
Wendt A (2005) Agency, teleology and the world state: a reply to Shannon. Eur J Int Relat 11(4):589–598.
Wolf AT (2009) A long-term view of water and security: international waters, national issues and regional tensions. In: Lipchin C, Sandler D, Cushman E (eds) The Jordan River and Dead Sea Basin. Conference Proceedings, Berlin, 2007 pp 3–19
Worsley P (2013) Marx and Marxism. Routledge
Zalewski P (2013) How Turkey went from 'zero problems' to zero friends. Foreign Policy. https://foreignpolicy.com/2013/08/22/how-turkey-went-from-zero-problems-to-zero-friends. Accessed 31 January 2023
Zarei M (2020) The water-energy-food nexus: a holistic approach for resource security in Iran, Iraq, and Turkey. Water-Energy Nexus 3:81–94

Chapter 3
Exploring the Sub-state–Supra-state Nexus in Transboundary Water Politics

Atal Ahmadzai

Abstract Theorizing transboundary water in International Relations (IR) theory explores the role of shared basins in shaping power dynamics between co-riparian states. The conflict-cooperation dichotomy expounds on shared water in a utilitarian context that can either cause conflict between or converge the interests of the riparian states. On the critical side, hydro-hegemony offers a basin-wide hegemon-subaltern dyadic based on state-centered power asymmetries and hierarchies. Contributing to the debate on exploring the political implications of shared basins, IR theory, however, failed to escape its simplistic and deterministic assumptions. Considering co-riparian states as unitary competitors that seek unitary interests, IR theory overlooked sub-national and global political imperatives of transboundary water. The theory also could not adequately examine the complex and multidirectional nature of power dynamics in shared rivers. By studying the roles of the Pakistan military and the World Bank in water governance in Pakistan, this chapter explores the sub-state—supra-state nexus in dominating transboundary water politics. The hegemonic control of Pakistan military serves group-based interests. By offering technical, financial, and conflict mitigation expertise, the World Bank advances its global agendas, as well as, enables the military's domination in the water sector.

Keywords Transboundary water · IR theory · Conflict-cooperation · Basin hegemons · Hydro hegemony · Water governance · Pakistan military · World Bank

3.1 Introduction

Water has a political character. In addition to its crucial role in the emergence of ancient social and political organizations, water shapes power structures among and within the more complex modern societies. Accordingly, the political character of water becomes more prominent and consequential when it transcends sovereign

A. Ahmadzai (✉)
St. Lawrence University, Canton, NY, USA
e-mail: ahmadzai.atal@gmail.com

K. Szálkai and M. Szalai (eds.), *Theorizing Transboundary Waters in International Relations*, Springer Water, https://doi.org/10.1007/978-3-031-43376-4_3

boundaries. From instigating the first war between ancient polities in the third millennium B.C. to facilitating the establishment of the first ever international organization among nation-states in the mid-nineteenth century (Campbell 1949), transboundary water has meddled with and shaped the strategic relationship between co-riparian states for a long time. In today's world of complex interdependencies, shared water shapes new power dynamics and relations at the sub-state and supra-state levels.

The market has aligned the interests of power circles at sub-state and supra-state levels. In many regions of the world, mainly in developing countries, water scarcity and the need for water resources development have transformed water into a market commodity, and for that reason, a profitable industry. Such profitability has facilitated synergies between the interests of sub-state actors, mainly the political-military establishments of riparian countries and those of global institutions and corporations, in commodifying water resources.

At the domestic level, securitizing water and its governance by elite sub-state actors serves non-unitary and group-based interests. At the international level, powerful supra-state entities, including international organizations, enable the commodification of water in the hands of elite sub-state actors through providing financial and technical assistance. In theorizing transboundary water, looking at such dynamics between actors at the sub-state and supra-state levels has remained overshadowed by the state-centered assumptions of the mainstream International Relations (IR) paradigms.

The political implications of transboundary waters are more complicated than expounded within the realm of IR theory, including the simplistic conflict-cooperation dichotomy between inter-riparian states or the basin-wide hegemon-subaltern dyadic. While the "conflict" thesis holds transboundary water as a geostrategic resource in a zero-sum power milieu, the liberal "cooperation" supposition looks at water as a mutually beneficial commodity for synergizing and converging competitors' interests towards a win–win solution. On the critical side, the hydro-hegemony thesis theorizes shared basins in a holistic context characterized by state-based power asymmetries and hierarchies that lead to the basin-based hegemony of the powerful riparian country.

Inspired by the overarching Eurocentric assumptions of IR theory, including unitary rationality and interest, these explanations mostly overlook the roles of sub- and supra-state actors in exploiting transboundary water resources for their respective group interests and agendas. Hence, the existing IR explanations of transboundary water have a prominent state-centered Western orientation, with limited applicability in non-Western contexts of statehood. These explanations fail to expound functional theses to draw a representative picture of the matter. While transboundary water has become a bottleneck issue and a source of conflict between actors at different levels, including but not limited to riparian states, it is crucial to reexamine these waters beyond narrowly and deterministically focusing on states as the primary and unitary actors in dealing with the issue.

This chapter critically examines the existing state-centered theories of transboundary water, scrutinizing the unexamined role of sub-state and supra-state entities in transboundary water politics. Sub-state actors are circles within states that exercise hegemonistic control over water through embedded institutional power. At the supra-state level, international institutions, global conglomerates, and multinational corporations with water sector portfolios shape politics on transboundary water. In synergy, the sub-state and supra-state actors usher institutional and discursive powers to monopolize interventions in the development and governance of water resources.

This chapter argues that unitary interests and competition do not drive cooperation or conflict among riparian states on transboundary water. On the contrary, water is a profit-based sector under the exclusionary control of a specific sub-state power elite that, in synch with global actors, pursues and materializes its group-specific interests. These interests define the milieu in which transboundary water is managed and utilized. In addition to institutional power, the sub-state actors use discursive power (nationalistic, developmental, and security discourses) to legitimize and consolidate their monopolies over water. This renders their political power and legitimacy to coordinate efforts with supra-state entities that rent them technical, financial, and construction services.

The chapter first surveys the existing IR theories of transboundary water, including the conflict-cooperation dichotomy and the hydro-hegemony thesis. It explores the fact that instead of being based on the ontological and epistemological tenets of critical theory, the overarching state-centered unitary assumptions of the IR paradigm lie at the foundation of the hydro-hegemonic thesis on transboundary waters. Next, the chapter discusses the dominating role of sub-state and international actors in the transboundary water sector. It presents the dominant role of Pakistan's military and the World Bank in the water sector as a case study to explore the sub-supra-state nexus in water governance in Pakistan and sheds light on securitizing water for consolidating the monopoly of the sub-state actors over water governance. The chapter concludes that it is the specific interests of these actors rather than those of the state that shape and drive transboundary water politics.

3.2 Transboundary Waters in IR Theory

Understanding the social and political aspects of water remains a crucial research area in various disciplines. These inquiries have contributed numerous analytical frameworks for exploring the social and political roles water plays in shaping and defining power structures among various societal actors (Wilson et al. 2019). As its social and political implications transcend political boundaries, transboundary water becomes a strategic resource, placing it within the scope of International Relations (IR).

Although it is claimed to be not older than a couple of decades (Furlong 2006), the history of theorizing transboundary water within the IR discipline goes back to the final years of World War II. However, compared to the limited amount of Cold War-era literature about the role of transboundary water in the relationships between inter-riparian states, grand theorizing of transboundary water within the IR discipline was only initiated after the end of the Cold War. By applying critical, structuralist, and neo-Marxist analytical frameworks, scholars in the post-Cold War era have formulated abstract theoretical schemes about transboundary water and the relationships between inter-riparian states. Reviewing the existing literature reveals a stark contrast between the Cold War and post-Cold War era theories on the subject matter.

During the first two decades of the Cold War, the focus on transboundary waters within IR was primarily on developing legal and administrative measures for joint and cooperative schemes. These included drafting international conventions for internationalizing and unifying legislation on transboundary water (UN 1952), proposing principles for international law on transboundary rivers (Eagleton 1955), recommending 'equitable apportionment' over the 'sovereign riparian rights' and 'prior appropriation' doctrines (Lepawsky 1963), and adopting the Helsinki Rules on the Uses of Water of International Rivers (The International Law Association 1966).

Inspired by the embedded liberalism of the post-WWII international political economy, the Western powers and the newly established international institutions attempted to create rule-based international approaches for transboundary waters. Although grand theorizing transboundary water did not attract IR scholars' attention, some international law experts and international institutions attempted to put into practice new principles and doctrines about the management and governance of shared basins.

At the time, the surge in new international legal instruments facilitated the dynamism for internationalizing transboundary rivers. For the newly independent countries, their colonial experience, however, prevented them from accepting any initiative of sharing sovereignty over the management of transboundary basins. Co-riparian countries in Asia, Africa, and South America opposed the internationalization of shared basins (The International Law Association 1966). They instead chose the territorialization of shared waters and establishing "their" waters under their exclusive jurisdictions, and the apportionment principle of international law provided legal grounds for the territorialization of transboundary waters.

For example, Israel in 1953 rejected the multi-purpose river basin approach proposed by the UN to settle the country's conflict with Jordan over the diversion of the upper Jordan rivers (Berber 1959). In South Asia, India and Pakistan could not agree on an integrated basin-wide cooperative approach for the conflicted Indus River basin. Instead, they settled the conflict by appropriating the river basin into six tributaries, and each co-riparian country established exclusive sovereignty over three of them (Teclaff 1967). In 1959, Sudan and Egypt reached an agreement on the Nile that divided the surplus between the two riparian states.

In promoting basin-wide technical and legal solutions for transboundary water, early Cold War liberal internationalism ignored the political imperative and implications of shared water. By having a declarative and prescriptive nature, this focus limited the scope of theorizing transboundary water in IR. Post-WWII embedded liberalism was a political project that, unlike its rival tradition of realism, did not evolve within rich ontological and epistemological contexts. The proposed solution of internationalizing the shared basin was a simplistic technical approach based on precedent exclusive to the Western countries' experience of internationalizing their transboundary rivers, such as the Rhine in Europe, the Columbia River in North America, or the Nile in North Africa.

The newly decolonized non-Western co-riparian states that had experienced political suppression and the looting or exploitation of their natural resources by the Western colonial powers were interested in the political aspect of shared basins. They were eager to consolidate authority and sovereignty over resources within their territories (Bourne 1962). They wanted "their water" to be under their own exclusive sovereignty. Exercising exclusive jurisdictional control over parts of transboundary rivers that flow within their respective territories was perceived as actual rather than presumptive sovereignty, and having authority over a spatially continuous domain realized their sovereignty (Conca 2006).

While the main scholarly focus on transboundary water during the Cold War was on the technical (legal and administrative) aspects, initial IR scholarship on water conflicts also emerged, which linked transboundary water with peace and wars, mainly in the Middle East (see Ionides 1953; Doherty 1965). While providing contextual material for the operationalization of the basin-wide legal and technical management approaches, these studies can be considered the anthesis of the "water war" thesis that emerged in the later years of the Cold War. In addition to scholars and practitioners, politicians also connected water with war and peace to advance their political agenda. By linking it with world peace (Worster 1992) or declaring it as *casus belli* (Schaaf and Fifield 1963; Brooks 2011), some political leaders securitized water during the Cold War.

Changes in the international strategic imperatives by the end of the Cold War led to a surge in reconceptualizing threats and security. Accordingly, transboundary water attracted new significance by defining it as a strategic resource that can either cause conflict or facilitate cooperation among co-riparian states. This unitary dichotomy overlooked the role and function of relevant sub-national and global actors in transboundary water and co-riparian relationships. As such, the "water conflict" and "water cooperation" understandings are not grand theories of transboundary water in IR but the extension of mainstream IR understandings of strategic resources, such as territory or extractive natural resources, in shaping inter-state relations. Each of these theses are briefly discussed below.

3.2.1 Water Wars

The Neo-Malthusian cry of "water wars" evolved within the changing strategic landscape at the end of the Cold War, when some scholars started looking at water as a strategic resource (see Cooley 1984; Falkenmark 1989; Starr 1991). The emergence of non-traditional and transnational threats and the broadening scope of security convinced some scholars to sound warnings about water wars.

Homer-Dixon (1994) asserted that water scarcity instigates aggression between co-riparian states. Gleick (1993), on the other hand, argued that transboundary water causes political confrontation that leads to dialog and negotiation, and Mandel (1992) revealed that transboundary water conflicts happen due to either territorial disputes among the co-riparian states or technological intervention in water (i.e. dams and flow diversion). While Frey (1993) concluded that various domestic and inter-riparian factors make water conflicts more likely at national and international levels, Chellaney (2013) claimed that water wars are already happening between riparian countries through political, economic, and diplomatic means.

In a broader sense, the New-Malthusian water wars thesis has remained popular within political discourses, policy circles, and academic debates. However, the thesis also prompted rigorous criticism of its underlying theoretical and empirical tenets. Empirically, there is a lack of precedent for co-riparian countries waging water war against each other. In terms of theory, the thesis is criticized as deterministic and overlooking mediating factors such as technological innovations, global interconnectedness, the substitutability of natural resources, and market mechanisms in preventing water wars (Gleditsch 1998). Instead of instigating violence, critics argue that transboundary water facilitates cooperation between co-riparian states.

3.2.2 Water Cooperation

With a Cornucopian orientation, the water cooperation thesis is embedded in neoliberal political and economic thought, debunking the causal relationship between transboundary water and conflict, and claiming that utilitarianism, market responsiveness, and technological efficiency constitute the underlying logic of cooperation among co-riparian states.

Haas (2002) argues that while water scarcity can cause violent intra-state conflicts, it seldom causes inter-state conflict. Likewise, Wolf (1998) and Yoffe and Wolf (1999) have found no historical precedent of water wars and concluded that waging water wars is strategically irrational and economically inviable for co-riparian countries. On the contrary, empirical data on shared basins across the world from 1948 to 1999 indicates that cooperation among riparian states far outweighed incidents of conflict and violence (Yoffe et al. 2003). Swain (2001), on the other hand, argues that co-riparian countries prefer legal arrangements on transboundary water, while Elhance (1999) asserts that riparian states seek cooperation regardless of the existing power asymmetry and hegemonic factors. Similarly, Gizelis and Wooden (2010) found that the institutional factors of co-riparian states shape their capacities to adapt to their freshwater needs without escalating into inter-state wars.

The conflict–cooperation dichotomy dominated the debate on the role of transboundary water in shaping international relations in the post-Cold War world. However, neither of the theses has been able to provide a comprehensive theory on transboundary water in IR. They fail to consider domestic and international institutional and structural factors and functions in comprehending the role of transboundary water in shaping the relationships between inter-riparian states. On the contrary, each thesis is based on traditional state and Western-centered IR assumptions and hence has limited applications elsewhere. The assumption of the state as a unitary actor overlooks the role of sub-state and supra-state actors in shaping the behavior of riparian states on transboundary water.

Aimed at overcoming these ontological and epistemological shortfalls of the conflict-cooperation dichotomy, some scholars, mainly within the critical tradition, have attempted to grand theorize transboundary water within IR. Below is a brief review of this scholarship.

3.3 Hydro-Hegemony

The conflict-cooperation dichotomy has failed to adequately theorize the complicated nature of transboundary water. The main challenge is that while "water wars" approach has a theoretical orientation, the cooperation thesis is empirically driven. Independently, they fail to provide a rigorous analytical framework to explain the role of transboundary water in shaping the relationships between inter-riparian states.

For the last few decades, some scholars, mainly of a critical orientation, have attempted to expound theories beyond the simplistic and deterministic conflict-cooperation dichotomy. Instead of state-centric and unitary assumptions, these explanations focus on different power structures that shape inter-riparian states' relationships at various levels. This strand of critical scholarship offers a more complex, interlinked, and hybrid understanding of the dynamics between co-riparian states, i.e. hydro-politics.

Hydro-politics theorizes the relationship between inter-riparian states in a context characterized as a complex where conflict and cooperation co-exist (see, Mirumachi and Allan 2007; Zeitoun 2007). In addition, it examines the complex and sophisticated nature of water politics at four distinct levels: water management, domestic water policy, interstate, and global water politics (Mollinga 2008). Within this literature, hydro-hegemony emerged as a promising thesis to critically explore inter-riparian states' relations in terms of power asymmetries. It defines hegemonic power in water affairs as the influence of the powerful riparian state exercised by various (material to normative) means (Zeitoun and Warner 2006). These means provide the hydro-hegemon with the power to dictate the rules of engagement in the basin.

In general, hegemony in a shared river basin is a relationship between the co-riparian states where the powerful state, otherwise legally equal, coercively or consensually maintain control through different means of power. Rather than physically occupying the basin by directly extending its control, the riparian hegemon uses its superiority either as guidance or in a dominating manner. Such efficient exertion of power in the shadow of hegemonic coercion ensures that the subaltern will not readily revolt (Warner and Zeitoun 2008). As such, any cooperation induced by a hydro-hegemon in a basin is unfair and ultimately perpetuates conflict (Zeitoun and Mirumachi 2008).

The hegemon can be a co-riparian state. It can also be an external power, for example, an international institution (Henwood and Funke 2002), or even a global initiative such as Integrated Water Resource Management (IWRM) (Warner and Zeitoun 2008), which can facilitate setting up a hegemonic rule in the basin through both conditionality (sticks) and funds (carrots). Therefore, hydro-politics, either at the basin or the global level, is mostly hegemonic and driven by power asymmetries.

3.3.1 Critiquing Hydro-Hegemony

In theorizing transboundary water politics, the hydro-hegemony thesis could not escape the influence of the mainstream IR paradigms. The role of co-riparian states as unitary actors is the assumption that provides an analytical foundation to the thesis in explaining the power dynamics in a river basin. This causes two issues that limit the application of the thesis.

First, considering co-riparian states unitary actors and asserting that national boundaries make water political (see, Jägerskog and Zeitoun 2009) is an oversimplified deduction. Water is political even within national borders, shaping power dynamics and structures among different societal actors. Similarly, water facilitates synergism between sub-national and global actors, such as the synergy between the local and global anti-large dam movements in the 1980s. By considering co-riparian states unitary actors, hydro-hegemony ignores the domestic (sub-national and community-based) and global political imperatives of water.

Furthermore, the distribution of water-based power and influence is both within societies and between countries (WCD 2000). This multipolarity of actors and multi-directional power dynamics become consequential during water scarcity, when decisions about access to and use of water involve actors at the intergovernmental, governmental, regional, community, and household levels and often become highly politicized (Rathgeber 1996).

The analytical paradox of hydro-hegemony is that while it criticizes the simplistic conflict-cooperation dichotomy as ignoring the complex nature of hydro-politics (see Zeitoun and Mirumachi 2008), it considers the state a unitary actor. Although Zeitoun and Warner (2006) applied a multi-level hydro-hegemony framework to explore the consenting and non-consenting victims of water deals between states, their framework fails to look at power asymmetries and hegemonies at the sub-national level. IN addition, while the main concern projected is exploitation among states, exploitation within states or state collusion against citizens is ignored (Furlong 2006).

Secondly, the unidirectional conceptualization of power has challenged the analytical robustness of hydro-hegemony. Most studies of hydro-hegemony look at power as a unidirectional flow of influence and control (coercive or consensual) between the co-riparian states. These studies primarily identify the state's economic and military capabilities as the locus of power among co-riparian states. This is a deterministic understanding of the complex and multidirectional nature of basin-wide power dynamics. Accordingly, hydro-hegemony holds other potential loci of power, such as topology, location, and class, which are secondary to state power. However, recent examples of interventions in transboundary river basins in North Africa and South Asia prove otherwise. The fact that Nepal and Ethiopia (which are relatively less powerful upper-riparian countries on the Ganges and the Nile) have built large dams dismisses the idea that power trumps topology. This suggests that focusing exclusively on the role of the basin hegemon in shaping hydro-politics is an inadequate explanation.

Methodologically, the case studies that inform the hydro-hegemony scholarship seem influential in terms of the deterministic conceptualization of power. Most of the conclusions are inferred from case studies in river basins that are confounded and conflicted by other security and geostrategic imperatives, such as the Jordan, Ganges, Euphrates, Tigris, and Nile basins (see, Zeitoun 2007; Zeitoun and Mirumachi 2008; Warner and Zeitoun 2008; Cascao 2008). The complex nature of co-riparian states' strategic politics in these basins can obscure other essential power dynamics at sub-state and supra-state levels. These studies do not incorporate the influential role of sub-state and global actors in shaping inter-riparian hydro-politics, and overlooking these actors costs the thesis critical scrutiny.

In addition, this scholarship has not been able to detach itself from mainstream IR paradigms' Western-centered and Western-dominant constructs and prescriptions. While proposing critical inquiry to explain complex co-riparian relationships, Mirumachi (2015) reconstructed the conflict-cooperation dichotomy into a continuum. Similarly, Jägerskog and Zeitoun (2009) prescribe transforming the power asymmetry in the river basins of the global South by either influence (facilitating positive-sum solutions or helping hegemons act as leaders rather than bullies) or by challenging the hegemons. Such prescriptions suggest consolidating and cementing the institutional, ideational, and power inequalities between the global North and South. By placing the "international community" in a position of fixing the power imbalances in the non-Western shared basins, they mentally and practically believe in the hegemony of the global North in imposing solutions.

Furthermore, Furlong (2006) argues that IR theory is weak in addressing critical issues of power and hegemony due to its "hydro-territoriality trap," which is manifested in state-centered analyses and the state's sovereignty and control over its water. However, the problem of the IR discipline in theorizing transboundary water is more profound than this. The problem is a global political problem, the hegemony of capitalism, which is imminent within transboundary water practices and discourses. This hegemony is operationalized by established, although unscrutinized relationships between sub-state and global actors within the realm of transboundary waters.

3.4 The Sub-state–Supra-state Nexus: The Case of the Pakistan Military and the World Bank

Pakistan is among the 36 most water-scarce countries in the world. It has limited surface water resources, depleting groundwater, and high population growth (Habib and Wahaj 2021). The increasing water insecurity in the country has transformed water into a strategic resource. The fact that the country's freshwater supply is heavily dependent on the Indus River Basin, a complex transboundary river system in Asia, further reinforces the strategic significance of water. Furthermore, the ever-increasing need for water resources development has made water a lucrative industry for investors. As such, driven by its strategic and economic significance, water governance has been centralized in the hands of sub-state elite circles in Pakistan.

Pakistan has a long history of centralized and elite-serving water governance (Nabeel 2021). This centralization is mainly in the hands of the military, which has established an effective monopoly and hegemonic control over the water sector (Chaudhry 2019). In addition, the profitability of water resources development also attracts supra-state entities, and by providing financial and technical assistance, these entities facilitate "merchandizing water" (Singh 2009). The World Bank, as a supra-state institution, has a long history of providing technical and financial support to the water sector in Pakistan. The roles of these actors will be discussed in the following section.

3.4.1 The Role of the Pakistan Military in Water Governance

The Pakistan military runs commercial entities worth tens of billions of USD (*Dawn* 2016). It is also the largest conglomerate of business entities carrying out mega construction and real estate development projects (UNDP 2020). More specifically, the military has a rich portfolio in the water sector. Through its organizations, such as the National Logistical Cell (NLC) and the Frontier Work Organization (FWO), the military undertakes critical water-related projects, including constructing dams and barrages, building irrigation canals, constructing water supply schemes, and doing land reclamation work (Pakistan Army 2023).

In addition to its active involvement in the water sector through various military-run private entities and conglomerates, the military controls water resources and hydropower development through its hegemonic control over the relevant public entities. The government-owned public utility of the Pakistan Water and Power Development Authority (WAPDA) illustrates the depth of the military in governing the water sector in Pakistan.

WAPDA was established in 1958 to centralize the development of the water and power infrastructure in the country. The original portfolio of the entity entails a wide array of services, including power, irrigation, water supply, and flood control. In recent years, since thermal power production was delegated to the private sector,

WAPDA's portfolio has refocused on the development of water resources and hydropower. The institution, however, has remained largely inefficient in delivering services, and Pakistan's new water policy sees an urgent need to upgrade WAPD's capacity in planning, designing, and implementing major hydropower projects (NWP 2018). One of the reasons behind its chronic inefficiency is the history of the Pakistan military in controlling the entity, either directly or indirectly.

During the long history of dictatorship in the country, WAPDA has mostly remained under the direct control of the military. For example, from the late 1990s until the mid-2000s, the Pakistan army deputed 30,000 military personnel to manage WAPDA (Chengappa 1999). Similarly, in times of civilian power, the military has had de facto control of WAPDA by assigning high-ranking retired military officers to lead and manage the public entity. This tradition goes back to the 1970s (Hussain 1988). Having control over WAPDA is to expand the military power base to civilian and public spheres, and the hierarchical and militaristic governance of the public entity comes with consequences in terms of inefficiency due to widespread corruption and the misappropriation of funds.

In recent years, massive corruption scandals have been reported in connection with WAPDA. During the 2013–2014 and 2017–2018 fiscal years, the Auditor General of Pakistan (AGP) exposed embezzlements, misappropriation, fraud, and other irregularities in the entity (*Down* 2015; Ahmadani 2018). In 2022, the National Accountability Bureau (NAB) accused the ex-chairman of WAPDA, a retired military general, of misappropriating USD 753 million. Under his watch, the entity commissioned mega hydro projects, including expanding the country's largest hydropower project, the Tarbela dam. In addition, giant corruption scandals were also reported related to three other mega hydro projects, namely the Dasu, Diamer-Basha, and Mohmand dams, which are part of the USD 60 billion China-Pakistan Economic Corridor (CPEC) flagship project (*The Economic Times* 2021; ANI 2022).

Although the military control of public entities is inherently prone to corruption, embezzlement of state funds, tax fraud, and even brutal coercive practices regarding workers (Goh and Muravska 2012), the Pakistan military has succeeded in expanding its hegemonic control over WAPDA. Generally, controlling public entities, including that of WAPDA, is aimed at increasing the power of the military within the public and the private sphere (Hussain 1988). The question, however, is how the military is effective in its control despite the widespread corruption and chronic inefficiency in water governance. Securitizing water and raising it to "high" or "national" politics has enabled the military to establish and maintain its control. More specifically, transboundary waters play a prominent role in the securitization of water and its governance in the country.

Pakistan's dependency on transboundary water, mainly on the Indus Water System (IWS), has shaped a nationwide perception of insecurity. India, the upper-riparian country in most rivers in the system, is perceived to have strategic control over the IWS. This perceived strategic disadvantage triggers different elements within Pakistan's political and military circles, including the military, extremist groups, and civilian institutions (Rigi and Warner 2020), which subsequently legitimizes securitizing water by linking it with sovereignty and a "national importance" that prescribes and justifies a militaristic, nationalized, and elite-centered water governance. Pakistan's new National Water Policy explicitly designates water resources and governance as a national responsibility (NWP 2018). Such an approach limits the policy-making authority of the civilian government (Williams 2018).

Some studies identify civilian, religious, or technical institutions behind securitizing water and its governance in Pakistan (see, Rigi and Warner 2020; Litfin 1999; Barthwal-Datta 2009). These arguments, however, overlook the role of the military in securitizing water and mobilizing other actors, including religious groups and the media, to socialize society with the constructed discourses about water. Through securitizing water and its governance, the military expands its hegemonic control in the country. In addition, securitization also helps the military to restrict and challenge the role of the civilian government in water governance and negotiating with co-riparian states.

In general, the Pakistan military, as a powerful sub-state entity, controls domestic policy making and foreign policy formulation (Tudor 2014; Rumi 2015). It also dominates public discourse on a range of issues (Ahmad and Ebert 2013). Similarly, the military exercises firm control on domestic and transboundary water issues. While such monopolies are justified based on national security and interest, they also have economic imperatives in that water development projects and schemes bring immense profit to the military. On the transboundary front, the uncompromising stance of the military about shared water with India indicates corporate interests (Rigi and Warner 2020).

Developing water infrastructure, including building large dams, is a profitable industry, and in South Asia, the emerging water insecurities can easily justify such interventions, although the pattern is more prevalent in Pakistan and India (Singh 2009). In Pakistan, however, and unlike in India, technical and financial limitations restrict the relevant actors' ability to realize many of their envisioned schemes. These limitations are mitigated by the role of supra-state entities, mainly the World Bank, in construction-led development in Pakistan's water sector.

3.4.2 The Role of the World Bank in Pakistan's Water

The World Bank provides countries with financial and technical support for development, and this portfolio includes funding water resources development and governance. As river basins are the sources of most freshwater supplies, the role of the World Bank in assisting countries to develop and manage their basins is prominent. However, its involvement has been cautious and limited in transboundary river basins, especially in conflicted ones. Conflicts between inter-riparian states over shared basins prevent the World Bank from intervening unless requested by the actors involved (Kirmani and Moigne 1997) and as long as it is compatible with the strategic goals of the Bank. Once it has established its footprint in a basin, the Bank sets up its institutional hegemony through conditionalities and funds. Such hegemony, however, cannot be established without collaborating with a sub-state actor that seeks its own hegemonic rule. The Bank's involvement in the water sector in Pakistan illustrates such a hegemonic rule in collaboration with the military of the country.

The World Bank has been heavily involved in the water sector in Pakistan since the early 1950s. In 1952, the Bank approved its first loan to Pakistan for a water project (Briscoe and Usman 2005). Subsequently, after long years of negotiation, it successfully facilitated transforming the country's dispute with India over the Indus Rivers basin. Signed in 1960, The Indus Waters Treaty (IWT) became the proudest intervention of the Bank. The World Bank transformed the Indus basin conflict between India and Pakistan through the apportionment principle, which divided the basin's water between the two riparian countries. Beside the two riparian beneficiaries, the IWT cleared the Bank's way toward establishing a hegemonic influence over the Indus waters in Pakistan. Ever since, the Bank has expanded its hegemonic rule in the basin through conditionality (sticks) and funds (carrots).

The treaty also served the World Bank by expanding its portfolio in Pakistan, which has systematically transformed into institutional hegemony over Pakistan's water sector. After facilitating the treaty, the Bank has initiated providing vital financial and technical assistance to the government of Pakistan in developing its water resources.

Based on its strategic goals, the World Bank has followed a hammer-and-nail approach to water resources development in Pakistan. Throughout its long engagement in the country, the Bank's lending has been mostly directed toward infrastructure development. This is compatible with its strategic logic and global agenda, which considers water resources and their development a tool for economic development. Water is an instrument of growth and a precondition for economic growth (Briscoe 2011), hence, for economic growth, countries have to increase their water storage capacity. Facilitating the IWT and resolving the Indus Basin conflict has helped the World Bank increase its engagement in the country and realize its strategic logic of infrastructure development.

In the 1950s and 1960s, the World Bank proactively promoted industrialization as an instrument of economic growth, mainly in the newly created states. Large industries were considered the engines of growth and modernity, and mega hydro infrastructure, including dams and reservoirs, was deemed crucial for providing energy to emerging industries. To realize this goal, the IWT was instrumental in granting the World Bank legal and political rationale to set up its institutional hegemony over the water sector in the country.

Soon after facilitating the IWT, the World Bank mobilized funds and financed mega hydro and irrigation projects in Pakistan. On average, post-IWT, 16.5% of the Bank's lending to Pakistan was in the water sector. In 1950–1960, the average annual lending of the World Bank to the water sector was around USD 580 million. This amount increased to USD 620 million in 1960–1970 (see Briscoe and Usman 2005). An overwhelming majority of these funds went to hydro infrastructure development. Under the Indus Basin Development Fund (IBDF) (1960–1967) and the Tarbela Development Fund (TDF), the bank lent most of its funds to infrastructure development. During this time, major reservoirs and hydropower projects, including the Tarbela dam on the Indus, the Mangle dam on the Jhelum, and the Chashma on the Indus river, were constructed.

While the World Bank was influential in shaping the domestic water development agenda of the country, their prescription did not move the country toward water security. Paradoxically, despite the increasing storage capacity in the country and having realized mega infrastructure projects, water insecurity in the country has worsened, and the country's per capita water availability has decreased systematically. In 1961, Pakistan's per capita total renewable water availability was 5,357 m^3. By 2019, it had reduced to 1,139 m^3 (FAO AQUASTAT).

Parallel with the neoliberal economic agenda initiated in the 1980s, the World Bank started scaling back funding for mega infrastructure projects. Instead, the Bank worked with Pakistan's elite circles to promote privatization in the water sector. Based on this new model, the elite circles were successful in privatizing elements of the water sector and relevant public utilities, and in the 1980s and 1990s, a shift towards privatization became visible. This pattern came with negative consequences in that access to water for domestic consumption and irrigation dropped between 1995 and 1999. In addition, privatization mainly benefited the economic and political elites, which further marginalized the poor in general as well as poor farmers in particular (Siegmann and Shezad 2006).

The neoliberal economic logic justifies privatization for efficiency, transparency, and participatory governance. These were not the fruits of privatizing some elements of the water sector in Pakistan. On the contrary, water governance has become more centralized through privatization. The military, aiming to reinforce its hegemonic control and the centralization of water governance, has privatized much of water resource development to its commercial entities. Funded by international donors, the centralization has eventually played into the hands of those who convert water into a national security issue (Ahmad et al. 1997). While promoting the principle of decentralization, the World Bank provided financial and technical support in creating a highly centralized, nationalized, and elite-serving water governance in Pakistan (Nabeel 2021).

The efficiency of the military institution in privatizing the sector has remained a positive indicator for the supra-state entities, including the World Bank, in implementing their strategic agenda. In general, while the initial capital-intensive large-scale infrastructure development and the subsequent neoliberal models needed uninterrupted implementation, they have remained counterproductive in bringing self-reliant changes, participatory intervention, and democratic decision-making, as these imported agendas neglect local needs (Ahmad et al. 1997). The sub-supra-state nexus is a win–win situation for both actors. On the one hand, the nexus is in the service of realizing the strategic goal of the supra-state entity, imposing neoliberal economic order. On the other hand, it safeguards the economic interests and hegemonic control of the sub-state actor in the country.

3.5 Conclusion

The dynamics between inter-riparian states in IR are theorized either as binary outcomes of conflict-cooperation or as a basin-wide hegemon-subaltern dyadic. These analyses consider co-riparian states unitary competitor actors seeking unitary interests. While they have different ontological and epistemological orientations, both of these explanations overlook the role of sub-state and supra-state actors in dominating transboundary water politics.

Unlike the existing theories, transboundary water politics in the international context is not driven by the binary dyadic of competition between co-riparian states. On the contrary, transboundary water is an exclusive profit-oriented industry driven by the interest of specific actors that operate at national, international, and global levels. At the national level, in many developing countries, the water sector is under the auspice of powerful elite institutions and establishments, including but not limited to the military. As the water sector contains lucrative enterprises such as hydropower and construction, these institutions establish exclusionary control and hegemony over these resources. In addition, institutional power provides the sub-state entities with instrumental means to establish a monopoly over deliberations on hydro-politics and transboundary water. However, this is not the only means of power, and these actors also strategically securitize water to ensure their exclusionary control over it. Through exercising discursive power, the domestic power elites justify their hegemony.

Globalization has intensified vertical and horizontal interdependencies between sub-state and supra-state actors within transboundary water politics. Unlike what is promoted by the conflict-cooperation dichotomy and the basin-wide hydro-hegemony thesis, such interdependencies are not based on unitary and state-centered decisions and interests. In contrast, local and global elite institutions' nexus synergizes and converges the interests of power circles that operate in sub-state and supra-state landscapes.

This nexus has transformed transboundary water into a profit-based landscape that strategically and intelligently exercises institutional and discursive power to establish a monopoly over hydro-politics at domestic and global levels. Understanding such contemporary politics goes beyond the scope of mainstream IR theories, which focus exclusively on states as the main, if not the only, involved actors. Similarly, the critical thesis of hydro-hegemony, by concentrating on river basin-based power asymmetries and dynamics, falls short of adeptly theorizing contemporary hydro-politics.

To critically analyze and theorize such politics, it is crucial to first identify the sources of power structures at sub-state and supra-state levels and subsequently explain their interests. This will lead to a critical comprehension of local and global power elites seeking group-based interests by exploiting people's wants and needs and the integrity of nature.

References

Ahmad I, Ebert H (2013) Will Pakistan's India policy under Sharif shift strategically? Strateg Anal 37(6):667–674
Ahmad I et al (1997) Water, power and people: a South Asian Manifesto on the politics and knowledge of water. Regional Centre for Strategic Studies, Colombo
Ahmadani A (2018) AGP unearths massive corruption in accounts of WAPDA. Ministry of Water Resources. Profit. https://profit.pakistantoday.com.pk/2018/05/10/agp-unearths-massive-corruption-in-accounts-of-wapda-ministry-of-water-resources. Accessed 2 February 2023
ANI (2022) Pakistan: controversy, scandal surrounds Diamer-Bhasha Dam on Indus. https://www.aninews.in/news/world/asia/pakistan-controversy-scandal-surrounds-diamer-bhasha-dam-on-indus20220917150942. Accessed 2 February 2023
Barthwal-Datta M (2009) Securitizing threats without the state: a case study of misgovernance as a security threat in Bangladesh. Rev Int Stud 35(2):277–300. https://doi.org/10.1017/S0260210509008523
Berber F (1959) Rivers in international law. Stevens & Sons Limited, London.
Bourne C (1962) International water law in a world of sovereign states: perspective for the future. In: Smith SC (ed) Water resources and economic development of the West. Report 10. Committee on the Economics of Resource Development of the Western Agricultural Economics Research Council, Portland, pp 121–125
Briscoe J (2011) Invited opinion interview: two decades at the center of world water policy. Water Policy 13(2):147–160
Briscoe J, Usman Q (2005) Pakistan's water economy: running dry—background papers. Washington, D.C. : World Bank Group. http://documents.worldbank.org/curated/en/278641468098987847/Pakistans-water-economy-running-dry-background-papers.
Brooks P (2011) The future of water wars. Foreign Policy. https://foreignpolicy.com/2011/05/05/the-future-of-water-wars. Accessed 2 February 2023
Campbell J (1949) Diplomacy on the Danube. Foreign Aff 27(2):315–327. https://doi.org/10.2307/20030183
Cascao A (2008) Ethiopia—challenges to Egyptian hegemony in the Nile basin. Water Policy 10(52):126–155. https://doi.org/10.2166/wp.2008.206
Chaudhry R (2019) "An island of excellence"? How the Pakistan military reflects on its presence in the development sector. ZEF Working Paper Series, No. 185, University of Bonn. https://www.econstor.eu/bitstream/10419/206975/1/1679240269.pdf. Accessed 2 February 2023
Chellaney B (2013) Water, peace, and war: confronting the global water crisis. Rowman & Littlefield, Lanham, MD
Chengappa BM (1999) Pakistan: military role in civil administration. Strateg Anal 23(2):299–312. https://doi.org/10.1080/09700169908455048
Conca K (2006) Governing water: contentious transnational politics and global institution building. The MIT Press, Cambridge
Cooley J (1984) The war over water. Foreign Policy 15(54):3–26. https://doi.org/10.2307/1148352
Dawn (2015) AGP finds Rs980bn irregularities in power sector. https://www.dawn.com/news/1208273. Accessed 2 February 2023
Dawn (2016) 50 commercial entities being run by the Armed Forces. https://www.dawn.com/news/1272211/50-commercial-entities-being-run-by-armed-forces. Accessed 2 February 2023
Doherty KB (1965) The Jordan waters conflict. Carnegie Endowment for Peace, Washington
Eagleton C (1955) The use of the waters of international rivers. Can Bar Rev 33(9):1018–1034
Elhance A (1999) Hydro-politics in the 3rd world: conflict and cooperation in international river basins. United States Institute of Peace Press, Washington, DC
Falkenmark M (1989) Fresh waters as a factor in strategic policy and action. In: Davis K et al (ed) Population and resources in a changing world: current readings. Stanford University, Stanford, pp 245–261

FAO. AQUASTAT database. Food and Agriculture Organization. http://www.fao.org/nr/water/aquastat/data. Accessed 2 February 2023

Frey FW (1993) The political context of conflict and cooperation over international river basins. Water Int 18(1):54–68. https://doi.org/10.1080/02508069308686151

Furlong K (2006) Hidden theories, troubled waters: International Relations, the 'territorial trap', and the Southern African Development Community's transboundary waters. Polit Geogr 25(4):438–458. https://doi.org/10.1016/j.polgeo.2005.12.008

Gizelis T, Wooden AE (2010) Water resources, institutions, & intrastate conflict. Polit Geogr 29(8):444–453. https://doi.org/10.1016/j.polgeo.2010.10.005

Gleditsch NP (1998) Armed conflict and the environment: a critique of the literature. J Peace Res 35(3):381–400. https://doi.org/10.1177/0022343398035003007

Gleick PH (1993) Water and conflict: freshwater resources conflict and international security. Int Secur 18(1):79–112. https://doi.org/10.2307/2539033

Goh K, Muravska J (2012) Military-owned businesses: corruption and risk reform. An initial review, with emphasis on the exploitation of natural resource assets. Transparency International UK. https://ti-defence.org/wp-content/uploads/2016/03/2012-01_MilitaryOwnedBusinesses.pdf. Accessed 2 February 2023

Haas PM (2002) Constructing environmental conflicts from resource scarcity. Glob Environ Polit 2(1):1–11. https://doi.org/10.1162/152638002317261436

Habib Z, Wahaj R (2021) Water availability, use, and challenges in Pakistan—water sector challenges in the Indus Basin and impact of climate change. FAO. https://www.fao.org/documents/card/en/c/cb0718en. Accessed 2 February 2023

Henwood R, Funke N (2002) Managing water in international river basins in Southern Africa: international relations or foreign policy. In: Turton A, Henwood R (eds) Hydropolitics in the developing world: a Southern African perspective. African Water Issues Research Unit, Pretoria, pp 177–186

Homer-Dixon TF (1994) Environmental scarcities and violent conflict: evidence from cases. Int Secur 19(1):5–40. https://doi.org/10.2307/2539147

Hussain A (1988) Strategic issues in Pakistan's economic policy. Progressive Publisher, Lahore

Ionides MG (1953) The disputed waters of Jordan. Middle East J 7(2):153–164

Jägerskog A, Zeitoun M (2009) Getting transboundary water right: theory and practice for effective cooperation. Report No. 25. SIWI, Stockholm

Kirmani S, Moigne G (1997) Fostering riparian cooperation in international river basins: The World Bank at its best in developing diplomacy. The World Bank Technical Paper No. 335. https://digitallibrary.un.org/record/195413?ln=en. Accessed 2 February 2023

Lepawsky A (1963) International development of river resources. R Inst Int Aff 39(4):533–550

Litfin KT (1999) Constructing environmental security and ecological interdependence. Glob Gov 5(3):359–377

Mandel R (1992) Sources of international river basin disputes. Confl Q 12(4):25–56

Mirumachi N (2015) Transboundary water politics in the developing world. Routledge, London and New York

Mirumachi N, Allan J (2007) Revisiting transboundary water governance: power, conflict cooperation and the political economy. In: Proceedings from CAIWA International Conference on Adaptive and Integrated Water Management: Coping with Scarcity, Basel, 1999, pp 12–15

Mollinga PP (2008) Water, politics and development: framing a political sociology of water resources management. Water Altern 1(1):7–23

Nabeel F (2021) Recasting Pakistan's water scarcity challenge. Atlantic Council. https://www.atlanticcouncil.org/wp-content/uploads/2021/10/Recasting-Pakistans-water-scarcity-challenge.pdf. Accessed 2 February 2023

National Water Policy (NWP) (2018) Ministry of Water Resources, Government of Pakistan. https://waterbeyondborders.net/wp-content/uploads/2018/07/Pakistan-National-Water-Policy-2018.pdf. Accessed 2 February 2023

Pakistan Army (2023) Infrastructure Development. https://pakistanarmy.gov.pk/Infrastructural-development.php. Accessed 2 February 2023

Rathgeber E (1996) Women, men, and water-resource management in Africa. In: Rathgeber E, Brooks D (eds) Water management in Africa and the Middle East: challenges and opportunities. IDRC, Ottawa, pp 49–56

Rigi H, Warner JF (2020) Two-level games on the transboundary river Indus: obstacles to cooperation. Water Policy 22(6):972–990. https://doi.org/10.2166/wp.2020.090

Rumi R (2015) Charting Pakistan's internal security policy. United States Institute of Peace. https://www.usip.org/sites/default/files/SR368-Charting-Pakistans-Internal-Security-Policy.pdf. Accessed 2 February 2023

Schaaf H, Fifield R (1963) The lower Mekong: challenge to cooperation in Southeast Asia. Van Nostrand, Princeton

Siegmann K, Shezad S (2006) Pakistan's water challenges: a human development perspective. Sustainable Development Policy Institute. https://sdpi.org/sdpiweb/publications/files/A105.pdf. Accessed 2 February 2023

Singh R (2009) Transboundary water politics and conflicts in South Asia: towards 'war or peace'. Centre For Democracy and Social Action. http://www.indiaenvironmentportal.org.in/files/water._Final.pdf. Accessed 2 February 2023

Starr J (1991) Water wars. Foreign Policy 22(82):17–36

Swain A (2001) Water wars: fact or fiction? Futures 33(8–9):769–781. https://doi.org/10.1016/S0016-3287(01)00018-0

Teclaff LA (1967) The river basin in history and law. Martinus Nijhoff, The Hague

The Economic Times (2021) China's concerns caused a change of guard at CPEC. https://economictimes.indiatimes.com/news/international/world-news/chinas-concerns-cause-a-change-of-guard-at-cpec/articleshow/85319348.cms?from=mdr. Accessed 2 February 2023

The International Law Association (1966) Helsinki rules. https://brill.com/view/journals/rpwl/3/4/article-p1_1.xml. Accessed 2 February 2023

Tudor M (2014) Shifting tides in South Asia: renewed hope in Pakistan? J Democr 25(2):105–118

UN (1952) Legal aspects of hydro-power development of rivers and lakes of common interest. UN Economic Commission for Europe Committee on Electric Power. https://digitallibrary.un.org/record/707526?ln=en. Accessed 2 February 2023

UNDP (2020) The three Ps of inequality: Power, People, and Policy. Pakistan National Human Development Report 2020. https://www.undp.org/sites/g/files/zskgke326/files/migration/pk/NHDR-Inequality-2020---Overview-Low-Res.pdf. Accessed 2 February 2023

Warner JF, Zeitoun M (2008) International Relations theory and water do mix: a response to Furlong's troubled waters, hydro-hegemony and international water relations. Polit Geogr 27(7):802–810. https://doi.org/10.1016/j.polgeo.2008.08.006

Williams J (2018) Stagnant rivers: transboundary water security in South and Southeast Asia. Water 10(12):1819. https://doi.org/10.3390/w10121819

Wilson N et al (2019) Re-theorizing politics in water governance. Water 11(7):1470. https://doi.org/10.3390/w11071470

Wolf A (1998) Conflict and cooperation along international waterways. Water Policy 1(2):251–265. https://doi.org/10.1016/S1366-7017(98)00019-1

World Commission on Dams (WCD) (2000) Dams and development: a new framework for decision-making. Earthscan Publications Ltd, London and Sterling, VA.

Worster D (1992) Rivers of empire: water, aridity, and the growth of the American West. Oxford University Press, Oxford

Yoffe SB, Wolf AT (1999) Water, conflict and cooperation: geographical perspectives. Camb Rev Int Aff 12(2):197–213. https://doi.org/10.1080/09557579908400256

Yoffe S, Wolf A, Giordano M (2003) Conflict and cooperation over international freshwater resources: indicators of basins at risk. J Am Water Resour Assoc 39(5):1109–1126. https://doi.org/10.1111/j.1752-1688.2003.tb03696.x

Zeitoun M (2007) The conflict vs. cooperation paradox: fighting over or sharing of Palestinian-Israeli groundwater? Water Int 32(1):105–120. https://doi.org/10.1080/02508060708691968

Zeitoun M, Mirumachi N (2008) Transboundary water interaction I: reconsidering conflict and cooperation. Int Environ Agreem: Polit Law Econ 8(4):297–316. https://doi.org/10.1007/s10784-008-9083-5

Zeitoun M, Warner J (2006) Hydro-hegemony: a framework for analysis of transboundary water conflicts. Water Policy 8(5):435–460. https://doi.org/10.2166/wp.2006.054

Chapter 4
Trust in Transboundary Water Cooperation: A Social Constructivist Approach

Bota Sharipova

Abstract Many practitioners and scholars engaged in transboundary water cooperation acknowledge that factors such as trust, emotions, beliefs, values, and personal chemistry play a role in transboundary cooperation and conflict. At the same time, the prominent International Relations theories of (neo)realism and neoliberal institutionalism have limited explanatory power in explaining (non-)cooperation on transboundary rivers. Taking the phenomena of trust and mistrust and their implications on transboundary cooperation as an entry point, this chapter argues that social constructivism as a theory of International Relations can open up new and more nuanced perspectives for a better understanding of the dynamics in transboundary water interactions. The theoretical discussions are anchored in the Aral Sea basin as a case study.

Keywords Trust · Neorealism · Neoliberal institutionalism · Social constructivism · Aral Sea Basin · Water cooperation

4.1 Introduction

Although the complexity of transboundary water conflict and cooperation poses a number of challenges for scholars who analyze them with the help of International Relations theories. At the same time, this approach allows for a diversity of theories, approaches, entry points, and methodologies to try to embrace this complexity. Currently, the prevailing academic literature that explains the links between states in their conflict-cooperation relations primarily focuses on the factors of geographic location (Warner and Zawahri 2012; Elhance 2000), power asymmetries (Dinar 2009; Zeitoun and Warner 2006; Cascão and Zeitoun 2013), the economic interdependence of riparian countries (Klaphake 2005), the broader political context (Hussein 2017; Zeitoun and Mirumachi 2008), and the legal and institutional set-up (Gerlak

B. Sharipova (✉)
IHE Delft Institute for Water Education, Delft, The Netherlands
e-mail: b.sharipova@un-ihe.org

K. Szálkai and M. Szalai (eds.), *Theorizing Transboundary Waters in International Relations*, Springer Water, https://doi.org/10.1007/978-3-031-43376-4_4

and Schmeier 2016; Kittikhoun and Schmeier 2021). Other factors receive significantly less attention in academia, such as trust, emotions, perceptions, sympathy, antipathy—sometimes referred to as affective factors.

There are many examples of where these factors play an important role in transboundary (non-)cooperation. For example, the personal antipathy of state officials hindering cooperation, shared traditions facilitating negotiations, mistrust in data provided by other riparian states, a fear of disadvantage in cooperative proposals, personal chemistry or sympathy making it possible to reach compromise.

These affective factors cannot be explained through a rationalistic stance, which assumes that states only act with the aim of maximizing their benefits. For example, states may prefer to cooperate even when unilateral actions and non-cooperation can potentially bring more benefits Similarly, proponents of institutionalist approaches to analyzing transboundary interactions find it difficult to explain non-cooperation in the basins where treaties were signed and where river basin organizations exist. Developing new approaches to analyzing transboundary water (non-)cooperation can thus bring into light new factors that have been mentioned but not theorized before, and disentangling these factors can open up new perspectives and enrich our understanding of transboundary conflict and cooperation.

This chapter focuses on trust as an affective factor between states that share transboundary rivers. Arguing that International Relations theories, more specifically (neo)realism and neoliberal institutionalism, can provide applicable but rather limited lenses to approach trust, I conclude that social constructivism can hold stronger explanatory power. The case of transboundary cooperation in the Aral Sea basin is used to illustrate the theoretical discussions and demonstrate the important role of trust and mistrust in international interactions on water.

The chapter has three aims: on one side, to contribute to the literature on utilizing International Relation theories to better understand transboundary cooperation. On another, to enrich the literature on trust by adding insights from a specific field of transboundary cooperation, and finally, to join forces with scholars who explore various ways of approaching the affective factors that influence international politics.

4.2 Trust in Transboundary Water Cooperation and Conflict

In the academic literature on transboundary cooperation and conflict, there seems to be a general consensus about the importance of the factor of trust. It is explicitly mentioned either as an entry point for the commencement of cooperation over rivers (Zhang 2019) or as a cornerstone, a maintenance mechanism of cooperation (Barua 2018; Zhang 2019). At the same time, mistrust, or lack of trust, is referred to as a barrier to cooperation in many basins of the world, for example, the Nile river basin (Cascão 2008), some rivers in South Asia (Biswas 2011), the Mekong river basin (Zhang 2019), and the Brahmaputra river basin (Barua 2018; Barua et al

2019). Although both concepts of trust and mistrust are used widely in the academic literature on transboundary cooperation, explicit theorizations seem absent.

There have been some attempts to utilize trust and mistrust as *explanans* in transboundary (non-)cooperation. Zhang (2019) adopts the conceptualization of trust from the IR field when analyzing cooperation, or rather non-cooperation, between Mekong countries, where the concept of mistrust is used to explain the non-cooperative attitude of the lower Mekong countries towards supposedly trustworthy upstream China. Taking a different angle, Lumosi et al. (2020) utilize the framework of trust and shared group identities from the management discipline to explain transboundary cooperation in the Zambezi basin.

A wide array of literature also focuses on trust-building measures in transboundary cooperation (Susskind and Islam 2012), although without necessarily linking it to IR theories. For example, trust is often mentioned as an outcome of third-party interventions in strengthening cooperation in transboundary basins, assuming that improved trust leads to better cooperation. For example, describing the influence of the UNECE Water Convention (Libert 2015; Libert and Lipponen 2012), and several articles about the Brahmaputra dialogues (Barua 2018; Barua et al. 2018, 2019) use improved trust as an indicator of the positive impact of a certain cooperation-building intervention. However, as a result of the lack of direct indicators or models of assessing trust, the word 'trust' is used rather frivolously when there is no clear evidence of improved cooperation between riparian states, or there is an implicit assumption that improved interpersonal trust between individuals from riparian states leads to better cooperation.

As it can be seen from the literature, despite the importance of the factors of trust and mistrust, there is still a lot of unclarity as to how it is embodied in practices of transboundary water cooperation and conflict, and how it influences the dynamics of the relations between states. Seeking to answer this kind of question requires an adequate theoretical and conceptual basis, a sound methodological approach, and rich empirical data. As a first step in this direction, the main aim of this chapter is to explore potential avenues of how (mis)trust could be viewed in transboundary water cooperation and conflict through the lens of different International Relations theories.

4.3 IR Theories, Transboundary Cooperation, and Trust

Although they are relatively well-studied in other disciplines, the phenomena of trust and mistrust pose challenges when examined in International Relations (Ruzicka and Keating 2015; Kydd 2005). In the international water domain, it is not only heads of states who are involved in transboundary cooperation but also the representatives of the relevant water and environment agencies, ministries of foreign affairs, NGOs, and individual experts. Relations between individuals, as well as those between states are messy, erratic, and they require the involvement of at least two parties.

This complex, non-linear, and dynamic constellation is difficult to trace, conceptualize, operationalize, and theorize. Trust, being a phenomenon that comes into play when there is an issue at stake, becomes a very important concept when it comes to the interdependencies and vulnerabilities of states in cooperation on transboundary rivers. Various theories of International Relations make implicit or more explicit assumptions about the role of trust in the relations between states. The following section introduces how trust is viewed through the lenses of the arguably most prominent IR theories that are used to explain the dynamics of transboundary interactions, i.e. (neo)realism, neoliberal institutionalism, and constructivism.

4.3.1 (Neo)Realism

Realism and neorealism have been among the dominant theories in International Relations for the last 60 years (Oelsner 2007). These two theories introduce pessimistic assumptions about trust by positioning mistrust at the heart of their reality, as a driver and justification of the relations between states. According to some realists, trust and other attributes of human interactions cannot even be considered in international relations (Gralnick 1988). By trusting another state, you put your state in a vulnerable position, which can be used by perceived enemies. Thus, according to the adherents of realism, states seek to eradicate vulnerability by mistrust and by accumulating power (Kydd 2005).

Realist arguments underpin the "water wars" narrative that was prominent during the 1990s (e.g. see Homer-Dixon 1994; Starr 1991; Swain 2001). Proponents of this approach focus on the physical scarcity of water as a natural resource, arguing that states often tend to want to gain power against riparian states to control limited water resources, given its limited and precious character, and they can even opt for military interventions.

Upstream countries tend to use their advantageous geographical position and threaten countries located on the lower stream, as Turkey did in the Euphrates basin (Warner 2008) or in the Jordan river basin (Lowi 1993). Downstream countries lacking the power to control water try to balance by gaining and exercising other forms of power in order to outperform upstream countries. Examples of powerful downstream countries are Egypt in the Blue Nile basin (in 1979 Egyptian President Anwar Sadat said that water was "the only matter that could take Egypt to war again") and Uzbekistan in the Aral Sea basin (in 2012 Uzbek President Islam Karimov warned about possible water wars in Central Asia because of water).

Implications of trust cannot explain cooperation over shared water resources through a (neo)realistic worldview because trusting relations are an anomaly. High levels of mistrust prevent cooperation, and even through cooperation, the level of mistrust can only be decreased slightly. The only way for cooperation to occur between states is through their willingness to create alliances with one another, with the aim to balance power against a more powerful state—in other words, when cooperation helps them gain more power. Nevertheless, cooperation in such cases is not based on trust but on rational choice, where states make rational calculations of costs and benefits, and they act in their own self-interest (Carlsnaes et al. 2013; Sharipova 2019).

4.3.2 Neoliberal Institutionalism

The concept of trust gains more support among the adherents of neoliberal institutionalism, who state that trust can influence relations between states. This theory argues that cooperation between states is not only possible, it is also beneficial, and establishing institutionalized cooperation reduces conflicts and competition (Jervis 1999). In managing transboundary water, neoliberal institutionalism arguments gain credence and power through examples of states that have managed to cooperate in generating win-win solutions. Although this give-and-take approach is also based on rational calculations, in contrast with the realists' deterministic worldview, states agree that having a war is costly. This shift in the global water cooperation discourse since the end of the twentieth century has marked a transition from the realistic to the neoliberal realm (Julien 2012). This transition underpins the rise of the institutionalist approach to transboundary cooperation, the main argument of which is that institutionalized commitment to cooperation expressed through signing basin or region-specific treaties, joining international conventions, and establishing river basin organizations (RBO) can minimize the risks of conflicts, and they can lead to better cooperation and win-win solutions (Elhance 2000; Gerlak and Schmeier 2016; Kittikhoun and Schmeier 2021).

Through this lens, the motives of states can be explained in terms of creating, joining, and being members of bilateral or multilateral organizations or some other collective management body to sustain the benefits of cooperation. It is not necessary that a state trust another state that is part of this organization, it should rather trust the institution. Although becoming part of the institution implies passing some power to the institution, states prefer to choose this way, trusting that the RBO will consider the interests of all parties and balance the needs of all member states. This viewpoint can be supported by multiple examples where an upstream country prefers to manage rivers jointly with downstream countries, for example, Guinea, Mali, Mauritania and Senegal on the Senegal river, or in the case of Canada and the United States over the management of the Columbia river. This cooperation allows riparian states to develop mutually beneficial forms of cooperation instead of exercising power and demonstrating independence in decisions as upstream states (Sharipova 2019).

Despite numerous cases where institutions have demonstrated effectiveness in addressing water conflicts (Kittikhoun and Schmeier 2021), some argue that institutions can only play a positive role in creating a regime that helps build trust, and this in turn can create an environment that is conducive to solving a specific water conflict (Haftendorn 2000).

4.3.3 Social Constructivism

The social constructivism theory of International Relations goes beyond the rational choice-based decision model attributed to realism and institutionalism, ontologically bounded together as "water rationality discourse" by Julien (2012), who states that the calculations of the benefits and costs cannot fully explain the decisions of the states. According to Alexander Wendt, one of the most prominent scholars of social constructivism, "it is through ideas that states ultimately relate to one another, … and these ideas help define who and what states are" (Wendt 1999: 372).

In contrast to schools of thought focused on power and self-interest, social constructivism views relations between states through individuals' lenses and viewpoints. Adherents of the constructivist worldview argue that International Relations are built or constructed upon human thoughts and the behavior of individuals, not solely by pre-defined physical factors or power. Emphasizing the role of beliefs, this theory assumes that international life is determined by the beliefs and expectations that states have about each other, thus, interests are also determined by ideas. Material conditions are not pre-defined and determined, they are shaped by identity, norms, values, knowledge, and beliefs (Wendt 1999).

Although its theory status is questioned, social constructivism can be viewed as the most accommodating theory for trust in terms of its focus on individuals (as agents) shaping reality (as constructs), including international politics. Governments or states, being social constructions, only exist in virtue of the practices that instantiate them. Without denying the importance of material factors, constructivists do not see them as the only determinants in cooperation or conflict between states, instead they leave space for ideational factors. Constructivism opens up perspectives on how these factors are created and thus constructed first in the human mind and how they are then used to shape reality. With that, there is a possibility to de-construct and re-construct reality (Wendt 1999). According to Wendt (1999), agents create social constructions, and social constructions create agents. The theory argues that this process of "crafting", or construction by agents and social constructions, is mutual, and it is constantly happening. Thus, a constructivist approach provides more space for understanding and analyzing trust and mistrust as social rather than material factors, and their influence on the relations between states.

According to the social approach to trust in IR, trust implies expectations of positive outcomes of the behavior of the other party (Kydd 2005). These positive expectations imply a certain understanding and knowledge about another party's behavior. According to Hopf (1998), knowledge about the identities, norms, values, and beliefs of another party helps reduce uncertainty and increases predictability – it is easier to decide whether to trust or not to trust. Here again, constructivism provides a much wider array for analysis. Scholars studying trust in International Relations in the past decade thus implicitly or explicitly follow constructivist theory.

When explaining transboundary water cooperation and trust through a constructivist lens, it is necessary to keep in mind that the value of water is not a given or constant, rather, its value is socially constructed, i.e. it is defined by the societies interacting with it. Water can be perceived as an economic good, a basic human need, a sacred liquid, the lifeblood of the planet. Based on the attitude to water that has been socially constructed by each of the parties involved, water does not necessarily become a resource to fight about, it rather encompasses the values society attaches to it. Thus, conflict or cooperation over water is not determined in advance, it emerges as a collective choice instead (Julien 2012).

Constructivism does not see cooperation over water between states as the only field of analysis. A constructivist perspective calls us to step away from centering water in international relations and pay attention to the contextual factors involved in water cooperation, such as the economic and political relations between the states. Julien (2012) argues that context is important because constructivism assumes that states with good cooperative relations in other aspects will be more willing to cooperate over water and vice versa – neighbors with limited overall cooperation would most likely hold back from water cooperation. Constructivists argue that the politics around water resources follows the overarching international relations between states, and thus the trust that was already gained in non-water sectors spills over to water.

Embedded in this theory, trust plays a significant role in water cooperation between states, becoming one of the factors that affect decisions whether to cooperate. All else being equal, states would prefer to cooperate over water with states that are trustworthy based on their previous experiences over others that are not trustworthy or with whom they have no previous experience of cooperation (Sharipova 2019).

4.4 (Non-)cooperation in the Aral Sea Basin—IR Theories

A number of scholars have studied (non-)cooperation in the Aral Sea basin (ASB), using various theoretical and conceptual standpoints. Below is an overview of the major theoretical trends in the explanations of (non-)cooperation and (mis)trust in the basin provided by realists and institutionalists, followed by a discussion of cases where the constructivist approach can have stronger explanatory power than the realist or institutionalist point of view.

The Aral Sea basin in Central Asia is shared between Kazakhstan, Kyrgyzstan, Tajikistan, Turkmenistan, Uzbekistan (all former USSR states), and Afghanistan. These five former Soviet countries of Central Asia share many transboundary rivers, the largest two of which are Amudarya and Syrdarya. Early after independence, the riparian states signed treaties and established a river basin organization for cooperation on water. However, despite the extensive support of international actors (Sehring et al. 2019), regional cooperation on transboundary rivers remains unsatisfactory. The development of hydropower potential in the upstream countries is considered harmful for the irrigational agriculture of the economically stronger downstream countries. This upstream-downstream controversy occurs amid the internal social, political, and economic challenges of the countries, further exaggerated by issues of water shortage, water quality, land degradation, the aging of the infrastructure, and the impacts of climate change, such as floods and droughts. To mention a few irreconcilable positions, an unsatisfactory legal-institutional framework and a lack of political will are mentioned as the causes of weak cooperation (Diebold and Sehring 2012; Menga 2017; Sehring and Ibatullin 2021; Krasznai 2018). Some scholars highlight the fact that mistrust, as an expectation of the misbehavior of riparian states, is one of the main reasons for weak cooperation in Central Asia (Wolf and Newton 2007; Abbink et al 2010; Murthy and Mendikulova 2017).

4.4.1 (Neo)realist Perspective

Predisposition to interstate water-related conflicts in Central Asia has long been studied in academia, especially during the early stages of states' independence. Lately it is mentioned mostly in alarmistic reports and newspapers (International Crisis Group 2002; BBC News 2016; Egamov 2019, Donnellon-May 2022; also see Peña-Ramos et al 2021). Geographic positions along the rivers, an unequal distribution of water and other natural resources (Elhance 1997; Spoor 1998), and the socio-economic and political consequences of the desiccation of the Aral Sea (Kotlyakov 1991; Glantz et al 1993; Zholdasova 1997) provide rich food for scientific discussion for scholars adhering to a deterministic view about the influence of material factors in IR. Studies on hydrohegemony reveal that there is no hydrohegemon in the basin (Wegerich 2008), as the states exercise different forms of power to balance out the inequalities that occur either naturally or due to economic, political, infrastructural, and other forms of power (Menga 2018). However, despite the gloomy perspectives, at least officially the states have committed to cooperation.

4.4.2 Neoliberal Institutionalist Perspective

Almost immediately after the collapse of the USSR in 1991, Central Asian countries signed two basin-wide treaties: the Agreement between Kazakhstan, Kyrgyzstan, Uzbekistan, Tajikistan and Turkmenistan on Cooperation in the Field of Joint Water Resources Management and the Conservation of Interstate Sources in 1992, and the Agreement between Kazakhstan, Kyrgyzstan, Uzbekistan, Tajikistan, and Turkmenistan on joint activities in addressing the crisis around the Aral Sea and the zone around it, improving the environment, and enhancing the social and economic development of the Aral Sea region in 1993. Scholars and practitioners working in the basin share the opinion that there is a need to revisit the agreements, as their provisions are currently not being implemented. Attempts to develop a new treaty have also remained unsuccessful so far.

A joint river basin organization, the International Fund for Saving the Aral Sea (IFAS) was established in 1992–1993, with the aim of managing the shared waters of the basin jointly. Although the positive role IFAS has played in preventing water conflicts and providing a platform for dialogue on regional water cooperation and coordination with donors has been acknowledged (Sehring and Ibatullin 2021), many scholars and practitioners criticize the organization for its low effectiveness, including its failure to facilitate discussions over contested issues. For example, in 2008–2016 the most debated issue in the basin occurred around the construction of the massive Rogun dam in upstream Tajikistan, which downstream Uzbekistan heavily contested. The Rogun issue was discussed at the highest political levels but not within IFAS. Other critics are concerned about low levels of coordination, biased IFAS bodies, and a lack of mandate and power (Krasznai 2019; Diebold and Sehring 2012; Menga 2018).

The institutionalization of cooperation in the ASB has received significant attention in academia regarding its legal aspects (de Chazournes 2006; Dukhovny 2004; Vinogradov and Langford 2001; Janusz-Pawletta 2018), institutional set-up (Weinthal 2002; Sehring and Ibatullin 2021; Janusz-Pawletta 2015), and international water law and global water conventions (Ziganshina 2014 and later; Libert and Lipponen 2012; Libert 2017). A number of articles reflect the variety of approaches taken to analyze the institutional aspect of transboundary cooperation; however, there is little space in these research projects to integrate and study mistrust, which was either explicitly or implicitly mentioned in the majority of them as an obstacle to better cooperation.

Despite the existence of a river basin organization in the Aral Sea basin, mistrust and low cooperation is indicated by the parallel initiatives of national governments, which have some overlapping functions with IFAS. For example, the UN Multi-Partner Human Security Trust Fund for the Aral Sea Region (MPHSTF) was established in Uzbekistan in 2020, after the President of Turkmenistan had presented a UN Special Program for the Aral Sea basin in 2019 (UN SPAS). Tajikistan, taking a leading role in promoting the water agenda worldwide under the UN umbrella, seems to follow its own path in the area of international cooperation in water. Although the openness of these initiatives to riparian states has been stated by the highest political leaders of these states, there seems to be little collaboration in these parallel international initiatives calling for cooperation. It could be argued that these initiatives are aiming to link individual states to players outside the region rather than promoting cooperation within the basin, or being represented globally as a basin as a whole through IFAS.

Putting aside the argument that a lack of hostile relationships and wars can be by itself considered a positive result for IFAS, the existence of the agreements and river basin organizations in the ASB has not strengthened trust among the states. What is more, in light of the recent autonomous initiatives of Tajikistan, Turkmenistan, Uzbekistan, and Kyrgyzstan to "freeze" their participation in IFAS in 2016, there seems to be little trust in the ability of IFAS to resolve regional water issues.

4.4.3 *Prospects for Trust Research Through Social Constructivism*

As we could see, the arguments of realists and institutionalists have limited explanatory power in explaining the co-existing dynamics of cooperation and non-cooperation in the Aral Sea basin. The question is then if constructivism can offer an applicable lens to analyze the dynamics of transboundary cooperation and contribute to the existing debates—below we present some cases where it can.

Scholars and water practitioners share the opinion that the year 2016 marks a new era of water thaw in transboundary cooperation in the Aral Sea basin. The new President of Uzbekistan, Shavkat Mirziyoyev, has opened up new prospects for Uzbek-Tajik cooperation, one of the foremost of which is bilateral cooperation in the water and energy sector. His predecessor, Islam Karimov, had problematic relations with Emomali Rahmon, the President of Tajikistan, which had negative implications for Uzbek-Tajik relations in many respects (Szálkai 2022; Fazendeiro 2017; Oxford Analytica 2017; Lemon 2016). With that, the Rogun case, which had prompted mistrust in Uzbekistan due to Tajikistan's potential power to control water and had for a long time hindered cooperation in the region, was solved virtually overnight. However, all other factors, except the Presidents, have remained unchanged.

While the benefits maximization arguments of realists or the commitments-based approach of institutionalists both fall short in explaining these changes, social constructivism, with its scrutiny of reality constructed in individuals' mind, offers an explanation of this change. According to the social constructivist worldview, the change can be explained through the high importance of interpersonal relations, such as personal bonding or animosity between heads of states in contexts where Presidents have the strongest decision-making powers, as is the case in Central Asian states.

Another case where mistrust can be observed in the structure of IFAS bodies is the rotation of the headquarter of the organization. The Executive Committee of IFAS (hereinafter EC IFAS) has to move every three years, as the member states have not been able to agree on a permanent location. With that, the Executive Committee has been located in Almaty (1993–1997), Tashkent (1997–1999), Ashgabat (1999–2002), Dushanbe (2003–2009), Almaty (2009–2012), Tashkent (2013–2016), Ashgabat (2017–2020), and Dushanbe (2020–2023). Every country is supposed to send two permanent representatives to EC IFAS, and these representatives have to move to a new country every three years with their families. Every three years, an institution has to start functioning from scratch with new staff in a new country, with the accompanying organizational, practical, and financial burdens. There are implications for projects that last longer than three years. In addition, the establishment of a new chairmanship, including its approval by all IFAS member countries, the nomination of the chairperson, the hiring of new staff, and the provision of facilities takes up to 1.5 years, with only 1.5 years remaining for actual work (Sehring and Ibatullin 2021).

Although rotating chairmanship is usual practice worldwide, this is not the case for the costly and time-consuming relocation process with each chairmanship, the procurement of the premises and technical equipment, and hiring new staff—a process that disturbs the normal functioning of the organization. The inefficiency of the process has been acknowledged on many occasions, but there is no willingness to change the situation due to a fear of the dominance of one country in the daily functioning of the organization. Following Julien's approach, this case demonstrates that perceptions about potential threats can be human-constructed. There are no material conditions that create a direct threat to one or another state, and there is no immediate power to gain. Still, the perceived risks, caused by mistrust and a fear of the dominance of one state in the regional institution, hinder agreement about the permanent location of the body and cause the inefficient performance of the organization.

These two examples illustrate the promising nature of social constructivism to provide fertile ground for further research on trust and mistrust in transboundary water cooperation through the lens of IR theories. Studies by Allouche (2005, 2020) about identity and water nationalism in Tajikistan, by Menga and Mirumachi (2016) and Menga (2018) about the role of ideational power in the region, have contributed to developing a more sophisticated understanding of how non-material factors influence interstate relations. Further research in the social constructivist vein can enrich our understanding of the complexities of transboundary water conflict and cooperation and the role of (mis)trust in ASB and other transboundary basins. Although national interests, power, and institutions might be the determining factors in the relations between states in some basins, this does not necessarily mean that these factors are equally important in all other transboundary basins. Social constructivism theory has the potential to challenge scholars and to encourage them to search for new theoretical, conceptual, and methodological tools to disentangle the intricacies of international interactions on water.

4.5 Conclusion

There is growing recognition by practitioners and scholars that affective factors, especially (mis)trust, play a role in transboundary conflict and cooperation. However, there are inadequate theoretical, conceptual, and methodological approaches to elucidate their implications. This chapter has provided an overview of the theoretical approaches that can be utilized to study trust and mistrust, namely (neo)realism, neoliberal institutionalism, and social constructivism. As the example of transboundary (non-)cooperation in the Aral Sea basin shows, realist and institutionalist arguments provide valuable explanations of the complex reality in the basin. However, their ontological standpoints – mistrust, national interests, and power accumulation as the driver of interactions in the (neo)realist case, and the regime and trust-building role of institutions in neoliberal institutionalism significantly limit their ability to embrace the multifaceted and multi-level phenomena of trust and mistrust. Social constructivism, with its focus on individuals' constructs, provides a much wider array for understanding (mis)trust and its implications in transboundary water conflict and cooperation.

With the focus on norms, identity, beliefs and values, constructivism makes it possible to single out the role of non-material factors such as trust in international relations. Steps have already been made toward taking individuals' perceptions as a unit of analysis in transboundary water cooperation, and future research on trust and the role of other affective factors can also follow this path. However, although focusing on individuals' perceptions is a promising avenue for further research, it also poses questions to scholars about the mechanisms and channels through which individuals' worldview can find its way to formulating foreign policy in cooperation on transboundary rivers. This is where transboundary cooperation on international rivers can provide a rich empirical subject for IR research.

Scholars researching trust and mistrust in IR through constructivist theory have to make a number of important decisions on how to approach it: which level of analysis to focus on, how to link various levels, whether the focus should be on practices and norms, relations, or beliefs, or perhaps all of these. social constructivism, in contrast with state-centered (neo)realism and institution-focused neoliberal institutionalism, provides a wide, almost unlimited variety of possible approaches to study transboundary cooperation and trust, and science can only win from this diversity.

References

Abbink K, Moller LC, O'Hara S (2010) Sources of mistrust: an experimental case study of a Central Asian water conflict. Environ Resource Econ 45(2):283–318

Allouche J (2005) Water nationalism: an explanation of the past and present conflicts in Central Asia, the Middle East and the Indian subcontinent? Doctoral dissertation, Université de Genève

Allouche J (2020) Nationalism, legitimacy and hegemony in transboundary water interactions. Water Altern 13(2):286–301

Barua A (2018) Water diplomacy as an approach to regional cooperation in South Asia: a case from the Brahmaputra Basin. J Hydrol 567:60–70

Barua A et al (2018) Powering or sharing water in the Brahmaputra River basin. Int J Water Resour Dev 34(5):829–843

Barua A et al (2019) Re-Interpreting cooperation in transboundary waters: bringing experiences from the Brahmaputra basin. Water 11(12):2589

BBC News (2016) Will Central Asia fight over water? October 25 www.bbc.com/news/magazine-37755985. Accessed 20 Jan 2023

Biswas AK (2011) Cooperation or conflict in transboundary water management: case study of South Asia. Hydrol Sci J 56(4):662–670

Carlsnaes W et al (2013) Handbook of International Relations. Sage, London

Cascão AE (2008) Ethiopia–challenges to Egyptian hegemony in the Nile Basin. Water Policy 10(S2):13–28

Cascão AE, Zeitoun M (2013) Power, hegemony and critical hydropolitics. In: Earle A et al (eds) Transboundary water management. Earthscan, London, pp 40–55

de Chazournes LB (2006) The Aral Sea Basin: legal and institutional aspects of governance. In: The multi-governance of water: four case studies. State University of New York Press, New York, pp 147–171

Diebold A, Sehring J (2012) Water unites: from the glaciers to the Aral Sea. Trescher Verlag, Berlin
Dinar S (2009) Scarcity and cooperation along international rivers. Global Environ Politics 9(1):109–135
Donnellon-May G (2022) Are 'water wars' coming to Asia? The Diplomat. https://thediplomat.com/2022/09/are-water-wars-coming-to-asia Accessed 8 Jan 2023
Dukhovny VA (2004) Future issues related to long-term agreements between states in the Aral Sea Basin. In: Bayarsaihan T and McKinney D (eds) Past experience and future challenges: cooperation in shared water resources in Central Asia, Asian Development Bank, pp 98–105
Egamov A (2019) Uzbekistan's impending water crisis. The Diplomat. https://thediplomat.com/2019/09/uzbekistans-impending-water-crisis Accessed 8 Jan 2023
Elhance AP (1997) Conflict and cooperation over water in the Aral Sea basin. Studies in Conflict and Terrorism 20(2):207–218
Elhance AP (2000) Hydropolitics: grounds for despair, reasons for hope. Int Negot 5(2):201–222
Fazendeiro BT (2017) Uzbekistan's foreign policy: the struggle for recognition and self-reliance under Karimov. Routledge, London and New York
Gerlak AK, Schmeier S (2016) River Basin organizations and the governance of transboundary watercourses. In: Conca K, Weinthal E (eds) The Oxford Handbook of water politics and policy. Oxford University Press, New York, pp 532–548
Glantz MH et al (1993) Tragedy in the Aral Sea basin: looking back to plan ahead? Glob Environ Chang 3(2):174–198
Gralnick A (1988) Trust, deterrence, realism, and nuclear omnicide. Polit Psychol 9(1):175–188
Haftendorn H (2000) Water and international conflict. Third World Q 21(1):51–68
Homer-Dixon TF (1994) Environmental scarcities and violent conflict: evidence from cases. Int Secur 19(1):5–40
Hopf T (1998) The promise of constructivism in International Relations theory. Int Secur 23(1):171–200
Hussein H (2017) Whose 'reality'? Discourses and hydropolitics along the Yarmouk River. Contemporary Levant 2(2):103–115
International Crisis Group (2002) Central Asia: water and conflict. https://www.crisisgroup.org/europe-central-asia/central-asia/uzbekistan/central-asia-water-and-conflict. Accessed 12 Jan 2023
Janusz-Pawletta B (2015) Current legal challenges to institutional governance of transboundary water resources in Central Asia and joint management arrangements. Environ Earth Sci 73(2):887–896
Janusz-Pawletta B (2018) Legal framework of interstate cooperation on water resources of Central Asia. In: Zhiltsov SS et al (eds) Water resources in Central Asia: international context. Springer, Cham, pp 25–39
Jervis R (1999) Realism, neoliberalism, and cooperation: understanding the debate. Int Secur 24(1):42–63
Julien F (2012) Hydropolitics is what societies make of it (or why we need a constructivist approach to the geopolitics of water). Int J Sustain Soc 4(1–2):45–71
Kittikhoun A, Schmeier S (eds) (2021) River Basin organizations in water diplomacy. Routledge, London and New York
Klaphake A (2005) Economic and political benefits of transboundary water cooperation. In: Value of water: different approaches in transboundary water management. Proceedings of the International Workshop, 10–11 March 2005, IHP-HWRP, Koblenz, pp 91–99
Kotlyakov VM (1991) The Aral Sea basin: a critical environmental zone. Environ Sci Policy Sustain Dev 33(1):4–38
Krasznai M (2018) Institutional cooperation on water resources management in Central Asia. In: Zhiltsov SS et al (eds) Water resources in Central Asia: international context. Springer, Cham, pp 41–60
Krasznai M (2019) Transboundary water management. In: Smakhtin V et al (eds) The Aral Sea Basin: water for sustainable development in Central Asia. Routledge, London, pp 122–135
Kydd A (2005) Trust and mistrust in International Relations. Princeton, Princeton University Press

Lemon E (2016) Signs of improving relations between Uzbekistan and Tajikistan but tensions remain. Central Asia-Caucasus Analyst. http://www.cacianalyst.org/publications/analytical-articles/item/13405-sings-of-improving-relations-between-uzbekistan-and-tajikistan-but-tensions-remain.html. Accessed 22 Jan 2023

Libert B (2015) The UNECE water convention and the development of transboundary cooperation in the Chu-Talas, Kura, Drin and Dniester River basins. Water Int 40(1):168–182

Libert B (2017) Application of the UNECE environmental conventions in Central Asia. In: Zhiltsov SS et al (eds) Water resources in Central Asia: international context. Springer, Cham, pp 81–101

Libert B, Lipponen A (2012) Challenges and opportunities for transboundary water cooperation in Central Asia: findings from UNECE's regional assessment and project work. Int J Water Resour Dev 28(3):565–576

Lowi M (1993) Water and power: the politics of a scarce resource in the Jordan River Basin. Cambridge University Press, New York

Lumosi CK et al (2020) Evaluating trust and shared group identities in emergent social learning processes in the Zambezi river basin. Humanit and Soc Sc Commun 7(1):1–13

Menga F (2018) Power and water in Central Asia. Routledge, New York

Menga F, Mirumachi N (2016) Fostering Tajik hydraulic development: examining the role of soft power in the case of the Rogun Dam. Water Alternatives 9(2):373–388

Murthy SL, Mendikulova F (2017) Water, conflict, and cooperation in Central Asia: the role of international law and diplomacy. Vermont J Environ Law 18(3):400–454

Oelsner A (2007) Friendship, mutual trust and the evolution of regional peace in the international system. Crit Rev Int Soc Pol Phil 10(2):257–279

Oxford Analytica (2017) Uzbek-Tajik flights offer hope of Central Asian thaw. Expert Briefings. https://doi.org/10.1108/OXAN-ES220237/full/html. Accessed 8 Jan 2023

Peña-Ramos JA et al (2021) Water conflicts in Central Asia: some recommendations on the non-conflictual use of water. Sustainability 13(6):3479

Ruzicka J, Keating VC (2015) Going global: trust research and international relations. J Trust Res 5(1):8–26

Sehring J et al (2019) International actors and initiatives for sustainable water management. In: Smakhtin V et al (eds) The Aral Sea Basin: water for sustainable development in Central Asia. Routledge, London, pp 155–175

Sehring J, Ibatullin S (2021) Prolonging or resolving water conflicts in Central Asia? The International Fund for Saving the Aral Sea. In: Kittikhoun A, Schmeier S (eds) River Basin organizations in water diplomacy. Routledge, London and New York, pp 222–241

Sharipova B (2019) Whose right is right? Searching for a common "right" for promoting trust and water-related data exchange in Central Asia. Master thesis submitted to Oregon State University, IHE Delft Institute for Water Education and University for Peace. https://ir.library.oregonstate.edu/concern/parent/7p88cp22m/file_sets/gq67jz25x. Accessed 8 Jan 2023

Spoor M (1998) The Aral Sea basin crisis: transition and environment in former Soviet Central Asia. Dev Chang 29(3):409–435

Starr JR (1991) Water wars. Foreign Policy 82:17–36

Susskind L, Islam S (2012) Water diplomacy: creating value and building trust in transboundary water negotiations. Sci Diplomacy 1(3):1–7

Swain A (2001) Water wars: fact or fiction? Futures 33(8–9):769–781

Szálkai K (2022) Central Asian water cooperation after COVID-19: could the impact of the pandemic challenge the achievements of the recent 'water thaw'? Research Paper/OSCE Academy in Bishkek. https://www.ssoar.info/ssoar/bitstream/handle/document/78156/ssoar-2022-szalkai-Central_Asian_Water_Cooperation_after.pdf. Accessed 24 Jan 2023

Vinogradov S, Langford VP (2001) Managing transboundary water resources in the Aral Sea Basin: in search of a solution. Int J Global Environ Issues 1(3–4):345–362

Warner J, Zawahri N (2012) Hegemony and asymmetry: multiple-chessboard games on transboundary rivers. Int Environ Agreements Politics, Law Econ 12(3):215–229

Warner JF (2008) Contested hydrohegemony: hydraulic control and security in Turkey. Water Altern 1(2):271–288

Wegerich K (2008) Blueprints for water user associations' accountability versus local reality: evidence from South Kazakhstan. Water Int 33(1):43–54

Weinthal E (2002) State making and environmental cooperation: linking domestic and international politics in Central Asia. MIT Press, Cambridge

Wendt A (1999) Social theory of international politics. Cambridge University Press, Cambridge

Wolf AT, Newton JT (2007) Case study of transboundary dispute resolution: Aral Sea. Department of Geosciences, Oregon State University, Corvallis

Zeitoun M, Mirumachi N (2008) Transboundary water interaction I: reconsidering conflict and cooperation. Int Environ Agreements Politics, Law and Econ 8(4):297–316

Zeitoun M, Warner J (2006) Hydro-hegemony: a framework for analysis of trans-boundary water conflicts. Water Policy 8(5):435–460

Zholdasova, I. (1997). Sturgeons and the Aral Sea ecological catastrophe. In: Birstein VJ et al (eds) Sturgeon biodiversity and conservation. Developments in environmental biology of fishes, vol 17. Springer, Dordrecht

Zhang L (2019) Trust crisis and building trust in transboundary water cooperation along the Lancang-Mekong River. In: Stewart MA, Coclanis PA (eds) Water and power. Environmental governance and strategies for sustainability in the Lower Mekong Basin. Springer, Cham, pp 235–251

Ziganshina D (2014) Promoting transboundary water security in the Aral Sea Basin through international law. Brill Nijhoff, Leiden

Part II
Conflict and Cooperation over Transboundary Waters

Chapter 5
Hydro-Hegemony in East and Southeast Asia: The Case of the Lancang/Mekong River

Zoltán Vörös

Abstract This chapter focuses on China and the Lancang/Mekong river within the Greater Mekong Subregion through the theory of hydro-hegemony to understand the transboundary water conflict in the region. Originating in the People's Republic of China and flowing through five ASEAN nations, the Lancang/Mekong river has become a new source of international water conflict. The upper stream country, China, is interested in exploiting the hydropower capacity of the river to supply the energy needs of its eastern urban territories, whereas the downstream, developing countries, and most importantly, the poor people within these states are dependent on the river's capacity to provide water for the agriculture as well as fish for their fishermen, while governments have also begun to investigate the possibility of hydropower utilization. China is expected to reshape the region and the river via the Lancang-Mekong Cooperation (LMC), the Belt and Road Initiative, and bilateral and multilateral agreements. Diverse trade agreements, infrastructure projects, and investments enable Beijing to exert direct influence over the fate of these countries in order to establish a secure border region. Through hydro-hegemony theory and the above-mentioned Chinese strategies created to control the water resources of the river, this chapter introduces the case of China and the Lancang/Mekong river, focusing on the Lancang cascade, an example of power asymmetry around a transboundary water issue, and analyzing China's behavior as an obvious hegemon in the region.

Keywords China · Southeast Asia · Lancang/Mekong · Water nationalism · Hydro-hegemony

Z. Vörös (✉)
University of Pécs, Pécs, Hungary
e-mail: voros.zoltan@pte.hu

K. Szálkai and M. Szalai (eds.), *Theorizing Transboundary Waters in International Relations*, Springer Water, https://doi.org/10.1007/978-3-031-43376-4_5

5.1 Introduction

Although a number of academics already refer to China as a hydro-hegemon (Rein 2016), the objective of the present research is to see how and by what means China can control and dominate the environmental, economic, and socio-political aspects of the Lancang/Mekong River, analyzing the Chinese hegemony over the river. In brief, the chapter focuses on China's (hydro-)hegemonic behavior in the river basin to see whether hydro-hegemony theory can explain the moves riparian states tend to make. The theoretical framework of hydro-hegemony was coined by Warner and later elaborated by him within the London Water Research Group together with Zeitoun (Rein 2016). This framework is used here to explore the relationship between power and water issues related to the Lancang/Mekong river, as well as examine how China exploits its status.

The Lancang/Mekong river is 4,800 kms long, originating in China and flowing through five Southeast Asian countries (Myanmar, Laos, Thailand, Cambodia, and Vietnam), within a region that is home to around 300 million people (Kákai et al. 2019). The upper reaches of the river are called Lancang Jiang in China. These upper reaches, flowing between cliffs and mountains, are particularly well suited to harnessing hydropower. The river enters Southeast Asia through China's Yunnan province, where Chinese energy demands are made evident by the construction of new dams and reservoirs: Yunnan, the 'home' of the Lancang Jiang, is also known as China's hydropower battery due to its enormous potential capacities: this section of the river in Yunnan Province has a technologically viable hydropower potential of up to 27,490 megawatts as a result of the province's intense water flow and considerable elevation range (1,780 m) (Deng et al. 2021: 2).

The river is not only the breadbasket of the region, a source of water and food for 70 million people, and the engine of the economy, it is also home to many large water projects, and China's role as an upper stream country is prominent in the exploitation of the river. An export-oriented economy since the 1980s, China's natural resources have been depleted over the past few decades, and the pollution from unregulated factories, which have made the country the second largest economy in the world, has pushed the environment and wildlife to the brink of destruction. The processes in China are always complex, and its dynamic development makes it difficult to assess the consequences in real time, but the "factory of the world" remains the most polluting state in the world, its hunger for energy is drawing in new territories, regions, and energy sources (including rivers), while its environmental protection and use of renewable energy is also at the forefront of global trends, as is the use of hydropower (Vörös 2020).

However, hydropower also has a downstream problem, which in the case of the Lancang/Mekong river is similar to that of many other rivers: countries downstream of rivers regulated by dams, power plants, and reservoirs face the potential negative consequences of reduced water flows, lost sediment, or in some cases, dams that are opened during the rainy season, thereby further raising water levels and facilitating floods.

5.2 Theoretical Background

Hydro-hegemony theory employs a realist view of hegemony and international relations regarding transboundary water conflicts. According to Zeitoun and Warner, theories by the early 2000s tended to minimize the significance of power in water conflicts and cooperation. They suggest that transboundary water conflicts lie somewhere between the non-existent water wars and cooperation, and that the lack of wars is an outcome of power imbalances rather than successful cooperations (Zeitoun and Warner 2006: 436). In addition, the assertion of control over transboundary water resources can occur through any number of power tactics, ranging from the use of military force at one end of the spectrum to the conclusion of unequal water treaties and the construction of prejudicial "truths" at the other (Selby 2007).

In the setting of highly unequal power relations, the strategy adopted by the hegemonic power determines the allocation of water resources to a great extent. In certain instances, hydro-hegemons embrace a guiding role and style of leadership, providing well-distributed advantages to all water users; in other instances, they develop unilateral and exploitative forms of water management, with serious negative implications for hegemonised nations and communities (Zeitoun and Warner 2006).

This study uses definitions by Zeitoun and Warner, and Rein (Zeitoun and Warner 2006; Rein 2016) to understand hegemony, hydro-hegemony, and power. According to the original concept of hydro-hegemony theory, Zeitoun and Warner differentiated (according to Selby, unsuccessfully [Selby 2007]) hegemony from dominance by saying that a "hegemon [..] is someone who guides the way, say a torchbearer in uncharted territory. As such, hegemony can be considered as leadership buttressed by authority. In contrast, dominance is defined as leadership buttressed by coercion" (Zeitoun and Warner 2006: 438). Coming from this angle, while hegemony can be both positive and negative through its practices, hydro-hegemony can also involve either "enlightened leadership" with beneficial water sharing or "oppressive domination" with some kind of oppression (Zeitoun and Warner 2006: 439).

Coming from this approach by Zeitoun and Warner, within transboundary water conditions we have to see how power is interpreted. Rein describes power as "a need for achieving desired outcomes in sharing a transboundary river by influencing other actors, e.g. other riparian states" (Rein 2016: 131). This definition aligns with Petersen-Perlman and Fischhendler's point, who argue that power has been misinterpreted in the water resource literature. According to them, hydro-hegemony theory builds on three deterministic factors that define transboundary interactions, namely military power, riparian position, and a state's technological potential to exploit the resource (Petersen-Perlman and Fischhendler 2018: 276). In other words, low-gradient, fertile floodplains downstream are easier to harness, while upstreamer processes require infrastructural work (Warner 2008: 275). However, Petersen-Perlman and Fischhendler believe that hegemonic vulnerabilities should not be neglected, and non-hegemons might also have opportunities to influence other riparian states, even the hegemon (Petersen-Perlman and Fischhendler 2018).

On the other hand, Cascão and Zeitoun offer a method to evaluate power through four dimensions (2010: 31–32): geographical power, material power, bargaining power, and ideational power. Geographical power reflects on the spatial context, the position of the riparian states, while material power deals dominantly with economic and military power. Bargaining power looks at how an actor can influence the ongoing processes for getting the desired outcomes, for example creating bilateral or multilateral agreements, or refusing to participate in such initiatives. With ideational power, Cascão and Zeitoun reflect on "the capacity of a riparian to impose and legitimise particular ideas and narratives" (Cascão and Zeitoun 2010: 32). As Rein summarizes it, the "most common method of this type of power is the usage of sanctioned discourse by delegitimising other types of discourse with its own hegemonic rhetoric, hiding necessary data, sharing ambiguous information, having better knowledge, stalling, securitising, using issue-exclusion and co-opting" (Rein 2016: 133).

This chapter looks at China as a hegemon in the river basin through the lens of these power-related dimensions and projects and initiatives, examining through the Lancang cascade how and by what means China is able to control and dominate the environmental, economic, and socio-political aspects of the Lancang/Mekong river. Lately the question of the Lancang/Mekong river has also been overpoliticized due to both the increasing global relevance of China and its changing behavior in the region, which is indicated by the appearance of external actors, especially the United States in the discourse around the river.

5.3 The Lancang Cascade

The Chinese government had already decided to construct a massive Mekong cascade on the upper Mekong in the 1980s, but the actual breakthrough occurred between 2000 and 2005, when the so-called West-East Electricity Transfer Project was initiated as part of the 10th five-year plan. The project is part of China's solution to the existing energy constraints in the country's eastern industrial and urban areas, moving electricity from the west, "in large part because most wind, coal, and hydro power plants are concentrated in the country's inland provinces" (Gibson 2013). According to Gibson, rather than addressing the causes of these water and energy imbalances via conservation and other demand management approaches, China's authorities are "feeding the beasts" (Gibson 2013) by pursuing such projects. The fact that the provinces receiving the energy are introduced in such a negative context (beasts are hardly likeable creatures) demonstrates the geopolitical layer of discussions over the river and the existence of extra-regional power interests. The beasts are China's seven recipient provinces: Beijing, Tianjin, Hebei, Shanghai, Zhejiang, Jiangsu, and Guangdong, which consume approximately 40% of the total electricity of the country (Gibson 2013). However, these provinces are also responsible for the economic growth of the country and provide home for a large proportion of the society.

As of now, eleven dams are in operation, with one more under construction, and Ganlanba in its final planning stages (Fig. 5.1).

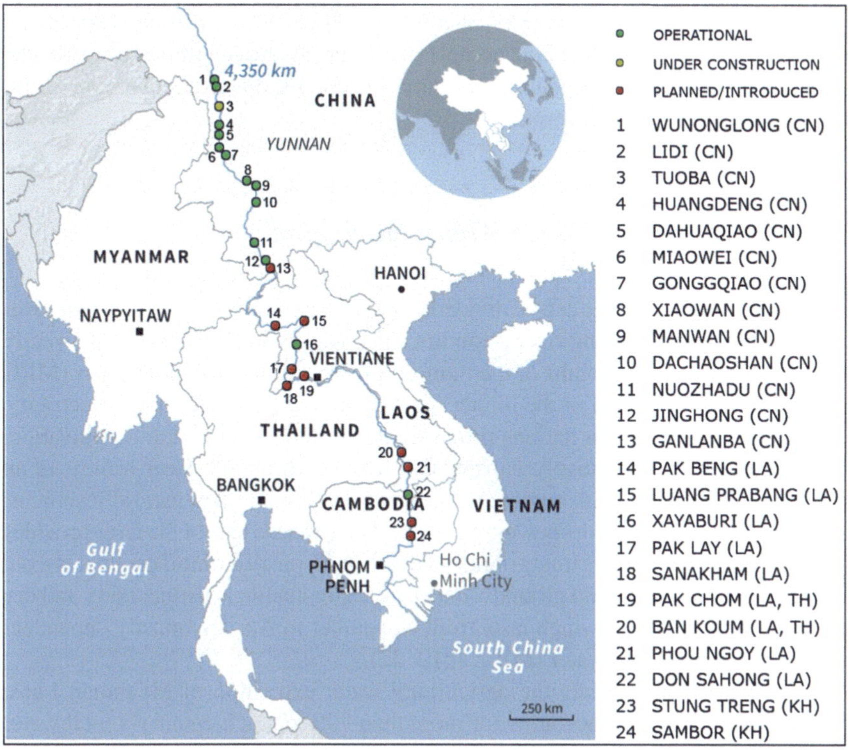

Fig. 5.1 Major Dams on the Lancang/Mekong River. Legend: CN—China, LA—Laos, TH—Thailand, KH—Cambodia. Based on data and locations by the Stimson Mekong Dam Monitor and Mekong Infrastructure Tracker

Huaneng Lancang River Hydropower Inc. operates all eleven dams, and the primary function of these facilities and the entire cascade is to generate power. Additional reasons for the projects included primarily drought and flood management, as well as navigation, albeit to a lesser extent on the Chinese side of the river.

The cascade and the reasoning behind building it highlight the goals of Beijing: solving power outages on their own, using clean, low-carbon, safe, and efficient hydroenergy while developing the underdeveloped western-inland territories, as well as providing projects for the construction companies. The World Commission on Dams remarked near the end of the twentieth century that from the 1930s to the 1970s, the building of major dams became associated with economic prosperity in the views of many. As dams were seen as emblems of modernization and humanity's capacity to control nature, their building proceeded significantly (WCD 2000: xxix).

The evaluation of the cascade is difficult, and not just because China's dam operation data are not completely transparent. In addition to criticism, there are also benefits, not only in the economic sphere, but also in terms of drought and flood control. In order to understand the complexity of the situation within the region, the fact that the usage of water resources in China and the five downstream nations differ significantly has to be understood.

5.3.1 Geographical Power Within the Region

The Lancang/Mekong river can be divided into two sections: the upper Lancang river basin and the lower Mekong river basin in China. Approximately 75% of the yearly flow occurs in just four months of the monsoon season, July through October (MRC 2019: 21). China, situated in the higher reaches, mostly generates hydroelectricity, while the five downstream nations primarily use water for agricultural cultivation. The river in China is relatively narrow, and it flows through a steep valley, in an underdeveloped region with cumbersome transportation and few natural resources, except for the abundant hydropower resources. The Chinese part of the river provides approximately 13.5% of the total volume of the river. Typically, runoff originates from precipitation, snowmelt, and groundwater. This region features distinct rainy and dry seasons. The dry season, which lasts from November to April, is mostly dependent on snowmelt and groundwater (MRC 2019: 22).

According to Deng et al., the agricultural water extraction of Myanmar, Laos, Thailand, and Cambodia accounts for more than 90% of each country's total water withdrawal, whereas Vietnam extracts 68% (Deng et al. 2021: 3).

The catchment areas and the population residing within the river basin are also indicators of the vulnerability of the lower-stream countries: Laos and Cambodia display the most dependence, with catchment areas comprising 97 and 86% of their respective territories, and 89 and 81% of their respective populations residing within the basin (Rein 2016: 138–139).

Problems also arise from the operation of Chinese dams: while storage dams (such as Nuozhadu, Xiaowan, and Huangdeng dams on the Lancang) might allow for longer-term planning of the river's resources, hydropeaking dams such as Jinghong can cause significant shifts for downstream communities. Hydropeaking takes place to satisfy local increases in energy demand, often during the day or during peak output hours. Downstream, the effect can disrupt the river's normal water flow and have an influence on its environment, impacting land animals, birds, and fish. (Roney 2022). In addition, the discharge procedure of hydropower stations during non-flood seasons is the most important human control component that can influence the water resource advantages (i.e. the benefits of agricultural irrigation) of the five downstream countries (Deng et al 2021: 3).

In conclusion, China's geographical location and relatively lower reliance on the river give it an advantage, and Beijing is clearly the geographic power in the region. Although not all states are significantly reliant on the river, and as previously mentioned, the Lancang only contributes 13.5% of the Mekong's total water yield, China is clearly the most geographically powerful nation in the region.

5.3.2 Material Power

From an economic and military standpoint, the capacities of China and the other riparian nations are almost incomparable, but two stories can help us put this aspect in context.

The China-Laos railway is one example that can help comprehend China's economic dominance. Beijing is expanding its regional presence through infrastructure projects in addition to its existing role as a significant investor and donor in the region. The 2021 opening of the railroad may be profitable for all parties and will strengthen the link between the two states. Starting with this project China plans to acquire a corridor through which physical connections with Thailand, Cambodia, Malaysia, and perhaps Singapore are possible. Given its own limited financial resources and infrastructure funds, Laos can also profit from these investments, which can transform it from a landlocked to a land-linked nation (Vörös and Somsack 2020).

Second, a Chinese initiative to actively push for regional security would underscore Beijing's dominance: as a somewhat surprising illustration of Chinese interest and presence, armed boats of the Chinese Border Guard arrive monthly to patrol the rivers all the way to the Thai border via Myanmar and Laos, which the Chinese say is to guarantee security, but many say is more of a show of force and intimidation. Joint patrol between these three countries was initiated in 2011, tackling smuggling and criminal offences (WSJ 2011).

These cases show how China approaches the region with the dual goals of integrating the region in an economic sense and creating regional security—and these goals are interconnected. Drug trade and the groups of organized crime[1] closely associated with security in the region are destabilizing forces, therefore Beijing thinks that combating narcotics and drug trafficking would also reduce security threats. Chinese officials are convinced that a decrease in security issues and an improvement in China's internal security will naturally accompany a rise in wealth in these regions (Baranyi et al. 2020), and the Chinese efforts are thus aligned with the geographical goals and visions of this emerging power (Vörös et al. 2022).

5.3.3 Bargaining Power

Based on China's geographical advantage and the vital operational data of the cascade, which might aid in the preparedness of the downstream nations, it is necessary for the parties to engage in an extended discourse. However, since China and the five nations have an unequal relationship, and downstream countries anticipate investments and further trade opportunities from China, it is difficult for them to convince China to alter its energy production tactics in the cascade.

[1] Referring to the so-called Golden Triangle (Myanmar, Laos, and Thailand), which was once the number one opium exporting region.

The likelihood of consultation is not evident. The Mekong River Commission (MRC), founded in 1995, might assist in coordinating river concerns when necessary; however, China (and Myanmar) are not members of this commission, making it impossible to confer on river matters. The Commission has recently emphasized that there is continued cooperation between China and its member nations, but it is frequently confined to flood control, the Commission's financial resources are limited, and China does not share future hydropower project plans.

However, there is another organization that could handle such conversations, despite its broader reach and goals. China proposed a regional cooperation framework to tackle the ongoing challenges at an ASEAN meeting in 2014, and the Lancang-Mekong Cooperation (LMC) was established in 2016. The fact that this framework was established and that the proposal was made at an ASEAN meeting indicate the Chinese considerations: becoming the driving force behind regional cooperation, using a multifaceted approach towards the region, interconnecting the diverse interests of the states, and avoiding a situation in which countries would criticize Beijing for the environmental consequences when they can expect to see benefits in other sectors. As the project website puts it, "LMC have identified the three cooperation pillars of political and security issues, economic and sustainable development, and social, cultural and people-to-people exchanges as well as the five key priority areas, namely, connectivity, production capacity, cross-border economic cooperation, water resources, agriculture and poverty reduction, and implemented many projects beneficial" (LMC 2017).

In the context of international and regional collaboration, China appears to reject one cooperation, the MRC, while supporting the one it founded, the LMC. The LMC is larger and by definition more significant for China, since it takes a comprehensive and multifaceted approach to the region, emphasizing security, which is vital for China. As Middleton and Allouche put it, LMC is a political initiative that represents a broader economic objective than only the Lancang-Mekong River. In an effort to expand China's influence, Beijing confronts existing institutions, including as the MRC and the GMS, over water resources (Middleton and Allouche 2016: 114). Since the goals of this organization are consistent with China's other regional visions and its economic and geopolitical vision, it is not surprising that Beijing has the upper hand in terms of bargaining power: the MRC cannot be effective without China, and the LMC, which is complemented by the instruments of Chinese foreign and economic policy, is attractive to all riparian states.

5.3.4 *Ideational Power*

To comprehend this subjective category, the effects of the cascade must be examined, as do the benefits, dangers, and challenges riparian states and international players, primarily the United States, are facing.

A number of repercussions of the cascade are indisputable, although they are subject to change depending on the perspective: consequences are connected to energy generation, navigability, and environmental effects influence silt levels and therefore biodiversity. Although the dams do not always run at full capacity, the generated electricity is a positive outcome, and navigability also presents economic gains; however, negative outcomes connected to these seemingly positive issues: in certain regions of Thailand and Cambodia, the catch of fish has reduced by half, allegedly due to lower water temperatures produced by the discharge of water from upstream reservoirs and the rapid hydrological variations induced by the opening and shutting of sluice gates to facilitate ship movement (Yeophantong 2014: 711). Another major ecological consequence in Thailand is the loss of breeding and spawning sites for fish and waterfowl, which have lost a safe, predator and hunter-free location to raise their young and develop (Bangkok Herald 2021), thus the cascade directly influences biodiversity and sustainability in a negative way.

Flood and drought control, and other economic consequences are not easy to define. The construction of huge dams and reservoirs on stretches of the river in China has in recent years led to water shortages or unexpected floods in the lower stretches of the river. According to a recent UN estimate, the sediment yield of the Mekong may already have fallen by 60% (Piman and Manish 2017), and the dams and reservoirs built by all the countries along the river without consulting each other are responsible for this outcome.

One example of Chinese influence and regional fears that is widely highlighted is the 2019–2021 water shortage in the lower reaches of the river, although it is not the only critical year for which a possible influence from China has been cited. Reflecting on the influence of China on the regional processes, the droughts of 2009–2010, 2012–2013, and 2015–2016, and a flash flood in December 2013 were already analyzed by the MRC in 2019, but jointly with the non-member China, and the so-called Great Mekong Floods of 2008 have also been linked to China by a number of downstream nations (Yeophantong 2014: 712).

Before examining the 2019–2021 extraordinary drought, the 2019 study emphasizes that the Lancang hydroelectric cascade has a positive influence on the flow and water level of the Mekong mainstream during the dry season and that the emergency water supplement from China during the dry season of 2015–2016 played a vital role in alleviating the drought on the Mekong river via the control of the Lancang cascade reservoirs (MRC 2019).

As it was mentioned above, China joined the investigation and helped put together this report, and coming from this, two totally different interpretations we could witness regarding the drought and flood management of the Lancang cascade—which we can introduce through the 2019(–2021) droughts. The first perspective is how China with joining MRC's investigation attempts to maintain control over defining the discourse—demonstrating their hegemony within this transboundary water issue. The other perspective is how US-based think tanks and experts attempt to seize the initiative demonstrates how riparian states can have or attempt to have power (Petersen-Perlman and Fischhendler 2018) by inviting the still-dominant global power into the discussion, similarly to how Vietnam and the Philippines acted within the South China Sea dispute.

The water level of the Mekong fluctuates periodically, with monsoon rainfall in the wet season and melting Himalayan snow boosting water levels, and dry seasons naturally decreasing them. This region in China experiences a wet season from May to October and a dry season from November to April. As we have seen, roughly 75% of the annual flow occurs in only four months of the monsoon season, July to October. Dams and reservoirs can influence this cyclicity in that the dams can aid in drought and flood management by refilling reservoirs during the wet season to prevent flooding, and releasing water during the dry season to prevent droughts. It is evident that the supplement water released from upstream hydropower stations is key to boosting downstream benefits, but it will also result in upstream power production loss; thus, the upstream hydroelectric stations must be aware of how much water they can supply and how much power they will lose to make water supplement collaboration more acceptable (Deng et al. 2021: 1).

Between June and November 2019, as well as in 2020 and 2021 during the wet season, the riverbed along the Thai-Lao border was almost completely dry in many areas, and it had dried up due to a lack of tributaries in others. A new estimate indicates that the extraordinary drought entered its fourth year in 2022 (Hunt 2022).

Who is accountable? China with its 11-dam-cascade or a potential drought in the Mekong basin? Based on climate data and satellite imagery, US climate research firm Eyes on Earth and US-backed think tank Stimson Center have determined that the Lancang basin of China has experienced above-average rainfall during these periods (June–November in 2019, 2020 and 2021), and according to their findings, China has effectively halted and held back water flow at an unprecedented rate (Eyler and Weatherby 2020). Although the predicted water flow from rainfall would have been available, the data reveals that the river water level has decreased. This was the first time in recorded history that Thailand did not experience monsoon season effects, and water levels did not rise (Eyler and Weatherby 2020). The research further indicates that the issue was exacerbated by the drought in Southeast Asia, noting that from 2000 to 2018, the Chinese region was somewhat wetter than average, whilst the Southeast Asian regions were drier than usual (Eyler and Weatherby 2020).

Possible causes of the 2019–2021 drought include China demonstrating its strength by filling its reservoirs (which is also occurring currently), although many have questioned the truthfulness of the report, including China. As a response, China has committed to releasing all hydrological information on the Lancang and agreed to participate in a fresh inquiry and report by the MRC through the LMC. This report, published in 2022, admitted that between 2019 and 2021 flows in the Mekong mainstream had dropped to their lowest levels in more than 60 years. However, overall the report found that the storages constructed in the Lancang basin are not the primary drivers of low flows to the downstream countries, these are rather caused by a combination of a worsening climate and unusually low rainfall (MRC 2022). Ironically, the report also recommended for the affected countries to coordinate their operational management, highlighting the still missing Chinese operational data: the Stimson Center revealed that China does offer hourly river-level data from a gauge at Jinghong below the Jinghong dam and another gauge at Man'an; however, these are river-level data and not operational data. As an illustration, Brian Eyler, director of the Southeast Asia program at the Stimson Center, cites an occurrence that took place in early October 2021, when data suggested that the level of the Jinghong reservoir had decreased, and the dam had discharged water. Satellite pictures show that the Jinghong dam had increased its reservoir volume, indicating that the rise in river level was caused by upstream dam releases that passed through the Jinghong dam (Roney 2022).

The fact that the Global Times has harshly criticized the US-based think tanks (Yuwei and Juecheng 2022), China is still not sharing operational data, and lately all MRC reports are put together jointly with the LMC show how China is gaining control over regional relations, and with referring on their ideational power, the country is using its data as a power tool.

5.4 Conclusions

Overall, hydraulic infrastructures can be used to wield power and undertake hegemonic and counterhegemonic projects. According to the original notion of hydro-hegemony, hegemons can establish a preferred form of interaction over transboundary waters. While Zeitoun and Warner talk about two possible scenarios, China is pursuing a strategy that alternatively employs leadership, a positive instrument, and dominance, a negative technique, and they have more experience with the latter.

This outcome is likely attributable to the political and internal political structure of the country, to its not quite transparent and one-sided decisions. As mentioned above, China's strategy is a multifaceted, regional perspective that attempts to resolve not only water challenges but also political, economic, and social dynamics. Here it is once again important to note that with this extended strategy towards downstream countries, through the LMC, China seeks to stabilize and secure the region to further its own social and economic objectives. The ASEAN-China Free Trade Area (ACFTA), for example, describes an economic interest: China was the first major power to sign a Free Trade Agreement with ASEAN, and as a consequence, trade and investment between the two countries have increased significantly. Since 2009, China has become ASEAN's largest commercial partner. In 2020, the group overtook the EU as China's largest trade partner. In 2020, bilateral commerce was valued at 731.9 billion USD, while foreign direct investment into the area exceeded 7.6 billion USD. From their extensive collaboration, connectivity has emerged as a major area, with both sides attempting to synchronize their respective agendas (Seth and Seah 2021: 3). It is also worth pointing out that, as highlighted above, the Lancang contributes about 13.5% of the volume of the Mekong River, so the vast majority of the water flow comes from tributaries and rainfall. In addition, due to the energy hunger and in order to regulate the river, the Southeast Asian states are also planning or building many dams—often with Chinese financial assistance, revealing a further layer of Chinese dominance.[2]

Although there are bodies (MRC, LMC) that could bring the parties to the table, cooperation concerns and misconceptions remain. Cooperation is not based on shared needs but on Chinese interests, as the history of the MRC and the formation of the LMC demonstrates. Beijing is willing to engage and discuss such procedures when it has an interest in them. Even with the sharing of current information on the flow of the river, cooperation is not flawless: the still-functioning MRC, despite its dwindling size and significance, coordinates, the China-dominated LMC assumes control, and numerous NGOs attempt to monitor the situation; nonetheless, Beijing continues to withhold information regarding the operation of its dams. While hydrological data is now available across the entire river, and China is willing to discuss and investigate important issues and problematic periods, operational data regarding the 11 Chinese dams is lacking. Even though these upstream dams have been able to contribute to the seasonal swings in demand in the nations downstream, the fluctuations have gotten more dramatic, and this pattern has continued through 2022.

It is evident that without transparency, it is impossible to distinguish between the effects of climate change and those of upstream dams, which can further complicate the effort to coordinate steps and efforts regarding additional hydropower infrastructures. Moreover, national interests (and energy hunger) seem to determine decisions, as more and more dams are announced by the countries concerned.

The fact that China does not share hydrological data with the MRC or with downstream countries, and that they continue to refuse to participate in a multilateral cooperation not dominated by China, only reinforces China's hegemonic and dominant role in the region. The river itself serves as a link between China and the downstream countries, not just through security-related measures but also through commercial relations, despite China's refusal to share vital operating data. While most recent NGO studies have been less critical of Beijing's hydropower activities, the government still has the ability to exert pressure on the Southeast Asian states by possibly controlling the water flow of the Mekong, thereby increasing its political and economic influence in the region.

---

[2] Altogether, there are 13 dams on the Mekong and 118 dams on the tributaries of the Mekong. As highlighted by Milton Osborne, in the lower Mekong nations, there are also hundreds of smaller dams used for irrigation, agricultural cultivation, fish harvesting, and water delivery. Together with sand dredging and a lack of the water levels necessary to wash the river clean, these dams have led to an increase in salinity (Hunt 2022).

References

Bangkok Herald (2021) Mekong dams inflicting untold damage of N. Thailand wildlife. https://bangkokherald.com/general/environment/mekong-dams-inflicting-untold-damage-on-n-thailand-wildlife. Accessed 15 Oct 2022

Baranyi TP et al (2020) A kínai Övezet és Út kezdeményezés – mítosz és valóság. KKI-elemzések, E-2020/98. https://kki.hu/wp-content/uploads/2020/12/KKIElemzesek.E-2020.98.pdf. Accessed 15 Oct 2022

Cascão AE, Zeitoun M (2010) Power, hegemony and critical hydropolitics. In: Earle A et al (eds) Transboundary water management. Earthscan, London, pp 27–42

Deng J et al (2021) Upper limit and power generation loss of water supplement from cascade hydropower stations to downstream under Lancang-Mekong cooperation. Water 13(20):2826

Eyler B, Weatherby C (2020) New evidence: how China turned off the tap on the Mekong River. The Stimson Center. https://stimson.org/2020/new-evidence-how-china-turned-off-the-mekong-tap. Accessed 15 Oct 2022

Gibson DT (2013) Mapping China's massive West-East electricity transfer project. Wilson Center. https://www.wilsoncenter.org/article/mapping-chinas-massive-west-east-electricity-transfer-project. Accessed 15 Oct 2022

Hunt L (2022) Mekong drought enters fourth year: worst conditions in 60 years. VOA. https://www.voanews.com/a/mekong-drought-enters-fourth-year-worst-conditions-in-60-years/6436938.html Accessed 15 Oct 2022

Kákai L et al (2019) 21. századi vízdiplomácia: a Mekong folyó esete. Nemzet és Biztonság: Biztonságpolitikai Szemle 12(3):104–124.

LMC (2017) 3+5 cooperation framework. http://www.lmcchina.org/eng/2017-12/14/content_41449855.html. Accessed 15 Oct 2022

Middleton C, Allouche J (2016) Watershed or powershed? Critical hydropolitics, China and the 'Lancang-Mekong Cooperation Framework.' Int Spectator 51(3):100–117

MRC (2022) Mekong low flow and drought conditions in 2019–2021: hydrological conditions in the Lower Mekong River Basin. https://www.mrcmekong.org/assets/Publications/Mekong-low-flow-and-drought-conditions-2019-2021df.pdf. Accessed 15 Oct 2022

MRC, IWHR, LMC Water Center, IWMI (2019) Hydrological impacts of the Lancang Hydropower Cascade on downstream extreme events. http://www.lmcwater.org.cn/authoritative_opinion/study/202009/P020200904567188882351.pdf. Accessed 15 Oct 2022

Petersen-Perlman JD, Fischhendler I (2018) The weakness of the strong: re-examining power in transboundary water dynamics. Int Environ Agreements: Polit Law Econ 18(2):275–294

Piman T, Manish S (2017) Case study on sediment in the Mekong River basin: current state and future trends. Project Report. Stockholm Environment Institute. https://www.sei.org/publications/sediment-mekong-river. Accessed 15 Oct 2022

Rein M (2016) Power asymmetry in the Mekong River Basin: the impact of hydro-hegemony on sharing transboundary water. Vienna J East Asian Stud 8:127–162

Roney T (2022) Lack of data on dam activity mars Mekong governance efforts. The Third Pole. https://www.thethirdpole.net/en/regional-cooperation/lack-data-dam-activity-mars-mekong-governance-efforts. Accessed 15 Oct 2022

Selby J (2007) Beyond hydro-hegemony: Gramsci, the national, and the trans-national. London: Occasional Paper No. 94 SOAS/KCL, Water Issues Group.

Seth FN, Seah S (2021) The ASEAN-China partnership: balancing merits and demerits. ISEAS Perspect 120: 1–14.

Spegele B (2011) China deploys patrols along Mekong. The Wall Street Journal. https://www.wsj.com/articles/SB10001424052970203430404577092092994816890 Accessed 15 Oct 2022

Vörös Z (2020) The water crises of China and its consequences on Southeast Asia. In: Kacziba P, Glied V (eds) Water stress: emerging challenges of global water scarcity. PTE, Pécs, pp 81–105.

Vörös Z et al (2022) Az Arany Háromszög felemelkedése és hanyatlása. Modern Geográfia 17(3):29–43

Vörös Z, Somsack P (2020) Laos and the Belt and Road Initiative: an interconnector helping the Chinese needs? Foreign Policy Rev 13:24–38
Warner JF (2008) Contested hydrohegemony: hydraulic control and security in Turkey. Water Altern 1(2):271–288
WCD (2000) Dams and development: a new framework for decision-making. The report of the World Commission on Dams. Earthscan, London.
Yeophantong P (2014) China's Lancang Dam Cascade and transnational activism in the Mekong Region: who's got the power? Asian Surv 54(4):700–724
Yuwei H, Juecheng Z (2022) GT investigates: who are the mouthpieces of US-led war of public opinion on "Chinese dams' threats" along Mekong River, and what are their typical methods? Global Times. https://www.globaltimes.cn/page/202203/1257016.shtml. Accessed 15 Oct 2022
Zeitoun M, Warner J (2006) Hydro-hegemony: a framework for analysis of trans-boundary water conflicts. Water Policy 8(5):435–460

Chapter 6
Reframing Transboundary Water Security

Seyedeh Zahra Ghoreishi, Hojjat Mianabadi, and Milad Jafari

Abstract Water security is one of the main goals of transboundary water policies. The current approaches to framing and achieving water security in transboundary river basins, however, are unsatisfactory and unsuccessful. Several factors interfering with establishing water security are overlooked by current approaches, including securitization and social dynamics, as well as neglecting environmental needs. This chapter aims to reframe the transboundary water security (TWS) perspective toward a more comprehensive and integrated approach to providing security. The foundations of TWS are investigated by scrutinizing what security is in International Relations and water security research. The chapter discusses who the security referent objects of transboundary arrangements are, and what actions and policies are needed to provide security. The theoretical investigations reveal the four security referent objects of state, development, society, and the ecosystem. Besides, the sustainability of referent objects is synthesized in a novel analytical framework. Nexus security, as a holistic approach which focuses on the complexity of referent objects and considers both negative and positive security is also suggested as a means to provide water security. Finally, the effectiveness and applicability of the developed framework is evaluated through practical investigations into the security of Afghanistan's transboundary river basins. The results indicate that Afghanistan has most frequently considered the state and development to be referent objects of security, through water nationalism, hydraulic missions, and populist rhetoric, making TWS less sustainable. In this regard, reframing water security and adopting a nexus security perspective

S. Z. Ghoreishi
Department of Irrigation and Reclamation Engineering, Faculty of Agricultural Engineering and Technology, College of Agriculture and Natural Resources, University of Tehran, Karaj, Alborz, Iran
e-mail: zghoreishy@ut.ac.ir

H. Mianabadi (✉) · M. Jafari
Department of Water Engineering and Management, Faculty of Agriculture, Tarbiat Modares University, Vardavard, Tehran, Iran
e-mail: hmianabadi@modares.ac.ir

M. Jafari
e-mail: milad.jafari@modares.ac.ir

K. Szálkai and M. Szalai (eds.), *Theorizing Transboundary Waters in International Relations*, Springer Water, https://doi.org/10.1007/978-3-031-43376-4_6

may assist policymakers in better understanding the ways to establish security in transboundary river basins.

Keywords Nexus Security · Transboundary waters · Security referent objects · Policy making · Afghanistan

6.1 Introduction

Addressing transboundary waters to move toward security is essential, as there are 310 transboundary rivers, 445 aquifers, and 1,600 lakes covering 47.1% of the Earth's land surface and 52% of the world's population (UNEP 2016; McCracken and Wolf 2019). The spatial extent of transboundary waters around the world links managing shared waters to intergovernmental arrangements and different security aspects, including socio-political, economic, national, and environmental (Thapliyal 2011). Transboundary water security (TWS) is thus a crucial constituent that affects the sustainability of water systems, while its lack jeopardizes socio-political and economic stability.

Water-related challenges and their multidimensional effects have been increasingly framed and analyzed through the concepts of water security (Grey and Sadoff 2007; Bakker 2012; Zeitoun et al. 2016; Sadoff et al. 2020) and transboundary water security (Albrecht et al. 2018; Ganoulis and Fried 2018; Williams 2018). Despite the growing interest in the literature, this chapter argues that the current water security frame and its understanding of transboundary basins are incomplete for three main reasons. First, water security is framed by objective indicators such as accessible water quantity and quality, and establishing water security means the absence of objective threats (Sadoff et al. 2020), thus often neglecting the intersubjectivity of water security (Mianabadi and Ghoreishi 2021). The second overlooked aspect is that adding the term "security" to water resources would invite the "securitization of water", putting transboundary water resource management decisions beyond the reach of normal politics (Zeitoun et al. 2016). The third aspect is failing to consider the nature of water systems and their interlinkages with other influential factors in managing water resources, such as food production (Hanjra and Qureshi 2010) and energy generation (Larsen and Drews 2019).

These gaps may promote an incomplete understanding of the current framing of water security, leading to an unsustainable, inequitable, and conflictive approach to using and exploiting water for both humans and the ecosystem. According to the current framing, transboundary water interactions may initiate water conflicts, resulting in water insecurity and destructive consequences for co-riparians (Jervis 1978; Dinar 2002). Therefore, the TWS policies that use the current framing may result in an unnoticed but predictable decrease in the water security index, the destabilization of socio-political relations, the emergence of conflicts, and consequently insecurity in transboundary basins (Rai et al. 2017; Srinivasan et al. 2017). Suprisingly the emergence of conflict and insecurity in transboundary basins may thus originate from the riparian policies following the establishment of TWS. Conflict-prone problems arising from the current water security frame make it necessary to reframe TWS.

Using the theoretical foundations of International Relations (IR) and guided by ongoing paradigms in water security, this chapter reframes TWS toward a more comprehensive approach to establishing sustainability and security in international water relations. By endeavoring to develop a novel analytical framework, more comprehensive policy pathways are demonstrated for establishing TWS. The developed framework explains the relationship between the influential aspects of defining TWS and the causal mechanisms that cause and/or hinder establishing security. It investigates two main questions: (i) TWS for whom? and (ii) TWS by whom?

The efficiency and applicability of the proposed framework is assessed in light of Afghanistan's current TWS circumstances. Water resources in Afghanistan are relatively abundant, with surface water availability exceeding 2,700 m^3/capita/year, well above the necessary 1,700 m^3/capita/year threshold (Thomas et al. 2016). Even though Afghanistan's geomorphological situation naturally provides it with a wealthy water status, there is a lack of water security (Asian Development Bank 2016). Afghanistan is among the least developed countries in the world, situated in a corridor of West Asia and Central Asia, and positioned upstream in four of its five river basins, which are transboundary. As a result, water policies in Afghanistan have inevitably impacted its co-riparians and may be a source of conflict and insecurity (Thomas et al. 2016). Evaluating and assessing the developed TWS framework by investigating the transboundary water policies of Afghanistan can be beneficial for understanding the water security conflict in the region, identifying policy pathways to TWS, and grounds for hope to ensure sustainability.

For the purpose of investigating the framework using transboundary water policies of Afghanistan, this chapter is based on an interpretative case study structure. Following this research design and theories of International Relations and water security research, the development of the framework is explained in Sect. 6.2. Subsequently, Sect. 6.3 is devoted to the analysis of the water policies of Afghanistan for a further assessment of the framework with the help of empirical evidence. Finally, Sect. 6.4 synthesizes the results, and a conclusion is presented in Sect. 6.5. For the research, traditional secondary sources have been supplemented by official documents, including action plans and publicly available reports.

6.2 Investigating the Foundations of TWS

Since there is a well-founded need for a comprehensive approach to water security, this chapter seeks to reframe the influential security components of TWS and thus requires a thorough theoretical foundation. The investigation of the foundations of TWS is guided by the paradigmatic literature of International Relations and water security research, and it examines three main components: (i) the security referent object, as an object or ideal that is threatened and needs to be protected (Buzan et al. 1998); (ii) threat factors; and (iii) policy pathways for establishing TWS. The deviations of water policies from a sustainable TWS are related to the relatively incomplete understanding of these components in the current framing. Subsequently, reframing these components is the main foundation of the proposed framework.

6.2.1 *The Referent Objects of TWS from the Perspective of International Relations*

Understanding the components that influence security and reframe TWS requires understanding the concept of security from the perspective of International Relations. Theoretical contributions in the field of International Relations are derived from various paradigmatic approaches to defining the slippery concept of security (Burke 2012; Zedner 2003). Although the literature presents a wide spectrum of security theories, the distinction between negative and positive security discourse is widely accepted, encouraging a broad recognition of security (Gjorv 2012; Roe 2008). A theoretical investigation of negative and positive security can thus serve as a foundation that leads to understanding the current TWS frame and identifying the complements required for reframing it.

Negative security is the treatment of security as a situation that one primarily wishes to avoid; it is the absence of threats (Gjorv 2012). The negative security definition is thus based on objective phenomena threatening national security and preservation. Positive security is a different approach based on enabling something desired, preferred, or interesting and subsequently recognizes security as an intersubjective concept (McSweeney 1984; Wendt 1999; Gjorv 2012; Huysmans 2002). This negative-positive paradigm reveals an understanding of insecurity that is due to objective or subjectively constructed threats. For analyzing and evaluating TWS, considering both objective and subjective criteria of water security is inevitable.

Regarding negative and positive security discourses, two theories have been proposed that reveal different security referent objects, threats, and policies envisioned (Waever et al. 1993): (i) securitization (Buzan et al. 1998) and (ii) societal security (Waever 2008). The theories of securitization and societal security indicate that state and social groups are two different security referent objects. By considering state or society as a security referent object, the threat factors and relevant actions for dealing with the threats could be revealed.

6.2.1.1 States as Security Referent Objects

According to securitization theory, security is defined as the absence of threats (Buzan et al. 1998). However, issues that become securitized may not always be existential threats. They are, instead, issues that securitizing actors, who transform normal problems into security issues, have successfully framed as threats (Waever 1995; Šulović

2010). Based on securitization, securitizing actors seek to implement "emergency measures" in response to perceived threats (Buzan et al. 1998). In other words, securitizing actors address self-interest issues by labeling them as threats to security, which allows them to act "beyond the law" (Bigo 2002). Thus, securitization is a management strategy for securitizing actors in political, economic, and environmental domains (Gharashi and Dehghani Firouzabadi 2013). In this vein, Waever et al. (1993) regard the state as the security referent object for multiple state-centric issues. Therefore, states can be TWS referent objects according to securitization theory, which could also be a management strategy for securitizing actors.

When states are regarded as security referent objects in transboundary basins, policies are altered to ensure the security of state-centric threats. Riparian states may rebrand something that is not a threat as a threat to take emergency measures that go beyond customary international water law to deal with perceived threats and pursue other self-interested goals. Governmental politicians and authorities may expressly become securitizing actors in managing transboundary waters. The state-centric goals could include providing legitimacy and acceptance to the state, as well as winning election races. To provide security for states, furthermore, managing transboundary waters could be represented as a security issue to construct a sense of patriotism and nationalism (nation-state building) while creating opportunities for self-interested and state-centric goal persuasion (Allouche 2005, 2019, 2020; Conker 2018; Menga 2016). Considering states as referent objects may encourage politicians to follow populist strategies to achieve state-centric security goals via transboundary waters.

Although transboundary water management may establish security for the state, from a securitization perspective, it may precede insecurity and a conflict-ridden atmosphere for the nation or co-riparians. If the states are merely considered as a referent object, TWS may be unsustainable; e.g., unilateral domestic development frequently results in conflict-prone transboundary relations (Allouche 2005). Furthermore, when states seek to securitize transboundary waters by instilling a sense of patriotism and water nationalism, the socio-cultural system may be rebranded by reshaping identity in an undesirable or self-interested manner. Subsequently, constructing identity in a basin country unit (BCU) entails erasing the previous collective identity of the entire basin and its inhabitants, who have a long history of coexistence. The destruction of collective identity may contribute to conflicts in transboundary river basins. Thus, instigating self-interested change processes in the socio-political structure of transboundary water arrangements in favor of a securitizing state may irreparably affect the sustainability of security and cause tangible or latent insecurity.

To sum up, considering states as security referent objects, the primary threatened factor is nation-states. States may often prefer to use populist measures for nation-state building or protecting the current circumstances through securitization. In fact, securitization and labeling water issues as threats may cause broader water-centric interests for states, such as water nationalism and patriotism, to take a role in establishing TWS. Despite that, securitization of transboundary waters will be for the fleeting state-centric purposes, in which goals of states may differ from the purposes of ruling regimes in political territories. These actions and policies may thus hinder the provision of stable and sustainable TWS. These approaches to TWS mean using water resources to accomplish state-centric goals and needs in BCUs, and they most frequently cause the unsustainability of hydropolitical relations, especially in the long term. For example, they can threaten riparians needs, construct new societal wants, augment developmental deficiencies, or jeopardize the ecosystem. For more clarity, policies for providing TWS for the states may threaten concerning desires of riparians and cause insecurity for the securitizing actors.

6.2.1.2 Social Groups as Security Referent Objects

In addition to states, social groups can also be security referent objects. This reference is derived from the theory of societal security. Societal security can be compromised when a community perceives something as a threat to important components of a social system, such as identity (Panić 2009). In this context, security means the absence of threats to identity, while threats to identity may instigate insecurity and conflict (Anderson 1983; Reimer et al. 2015). In comparison, McSweeney (1998) argues that societal security from a negative perspective is purely objective, so the intersubjective nature of security derived from positive discourse is overlooked. Subjectively, societal security is associated with a society's tolerance for dealing with threats via, for example, ideological coefficients (Eftekhari 2013), which are related to changes in society's "wants" and "belonging". Wants are defined as needs based on values, norms, and identity, while belonging is a collection of objective and subjective responses to wants. Therefore, in addition to identity, ideology is another security component that indicates the emergence of insecurity when the gap between wants and belonging increases (Bubandt 2005; Gjorv 2012; Roe 2008).

By focusing on identity and ideology in reframing TWS, the stumbling blocks to its establishment become more clear. When society is regarded as the referent object in transboundary basins, identity formation and alterity can pose a threat to TWS. Identity determines which riparians are friends of the state and which are enemies (Wendt 1999), which may thus shape riparian relations toward fluctuations of conflict and cooperation and may impact water interactions and TWS. For instance, introducing the assumption that conflicts over transboundary waters are not always rational, Kalpakian (2000) shows that in the Nile, Tigris-Euphrates, and Indus river basins, what matters is identity. Identity shapes people's attitudes and creates "others" over shared waters (Kalpakian 2000). Furthermore, Szalkai (2012) demonstrates how the meaning of the Amu Darya river basin for Uzbekistan and Tajikistan gradually changed from cooperative to conflictive when the two states became independent of the Soviet Union, ceased to cooperate, and eventually became each other's rival. Thus, the identity of riparians is important in TWS, and moving toward riparian state solidarity could benefit TWS in this regard.

Moreover, the fluctuation of wants and belongings is another security threat in transboundary basins based on societal security. For instance, failure to fulfill wants may jeopardize TWS, while aligning with belongings could result in the establishment of TWS. In this vein, riparians may seek to supply more water resources through conveyance projects or by increasing reservoir capacity, but contrary to conventional expectations, these fail to provide security in transboundary basins due to the increasing wants and needs. This problem can also be explained by the "fixes that backfire" archetype (Bahaddin et al. 2018), where fixing the symptoms and increasing belongings, such as water scarcity, is effective in the short term but intensifies the wants. Considering society as a TWS reference, the increase in belongings does not lead to security but fosters a gap between wants and belongings. Hence, demand management may be more constructive, and supply management policies may intermittently fail to guarantee sustainability and robustness.

6.2.2 The Referent Objects of TWS from the Perspective of Water Security Research

After this brief introduction to the main discussions on security in the field of IR, investigating the concept of TWS in water security research is critical to unpacking the TWS components and referent objects in order to reframe the current understanding of this issue, which requires a consideration of the views and evolutions present in the field of water security research. Although there are numerous definitions of water security, UN-Water (2013), for instance, advance the well-known definition of the "capacity of a population to safeguard sustainable access to adequate quantities of acceptable quality water for sustaining livelihoods, human well-being, and socioeconomic development, for ensuring protection against waterborne pollution and water-related disasters, and for preserving ecosystems in a climate of peace and political stability." This definition reveals two security referent objects in transboundary basins: (i) development, and (ii) the ecosystem.

6.2.2.1 Development as a Security Referent Object

Development is intricately tied to security, and it is a prerequisite of security establishment (Sayar 2012). It is also vital in water resource management and other sectors in terms of improving socioeconomic conditions. Both establishing security and improving socio-economic conditions make development one of the main issues addressed in the strategies, policies, and programming of countries. Development is thus linked to improving flood and drought management, adaptation to climate change, industrial activities, urban/rural sustainability, and systems of agriculture and energy production. Hence, establishing security in each sector is as important as TWS, with intricate connections among them (Sayar 2012). For instance, to provide food and energy security, water demand may increase and challenge TWS. This cross-sectoral interaction makes the water-food-energy nexus prominent in policymaking to establish security. Thus, establishing water security in transboundary basins is unattainable without development, and there is no development without providing cross-sectoral security (Sayar 2012). Accordingly, one of the main policies riparians adopt is development to achieve security in transboundary basins.

Development is the most serious environmental problem (Gleditsch 2003), especially in regions that face aridity, population growth, and protracted conflicts (Dinar 2003). When development is considered a TWS referent object, the effectiveness of actions and policies may become problematic. Hence, the policies of riparian countries may turn to the unilateral development of transboundary waters to establish security (Nagheeby and Warner 2018). For instance, development grants for hydraulic missions (damming with height infrastructures, water conveyance projects, and groundwater abstractions) may be spent to capture more water to overcome domestic socioeconomic issues rather than multilateral attainment of TWS (Molle et al. 2009; Hussein et al. 2020). The development policies and strategies of a BCU may ignore the importance of the joint management of transboundary waters, which will inevitably affect co-riparians (Wolf 1995). This is while unilateral water utilization may foster competitive behaviors and incompatibilities between riparians and, consequently, emerge conflictive challenges to sustainability of TWS. Riparians may subsequently unilaterally pursue development policies with the illusion of moving toward TWS.

6.2.2.2 The Ecosystem as a Security Referent Object

In addition to development, the ecosystem can also be recognized as a referent object of TWS based on water security research. According to the Millennium Ecosystem Assessment (2005), each water ecosystem provides numerous benefits to human society, including water, aquatic products, and hydropower. Moreover, the ecosystem regulates water purification, flood and drought control, and it is responsible for climate-induced events. Ecosystems pose nonmaterial benefits ranging from aesthetic or spiritual enrichment to recreation, scientific research, and educational activities (Millennium Ecosystem Assessment 2005). Ecosystem services connect human and water systems and provide a viable policymaking basis for establishing TWS. In this regard, riparians have an obligation to cooperate to protect the ecosystem due to Article 20 of the 1997 Watercourses Convention, which states that "watercourse states shall, individually and, where appropriate, jointly, protect and preserve the ecosystems of international watercourses". Furthermore, in July 2022, the U.N. General Assembly declared universal human rights access to a clean and healthy environment (UNGA 2022). This crucial role of ecosystems necessitates to consider them as TWS referent objects.

By recognizing ecosystems as a security referent object, the definition of security and insecurity changes. Threats against the security of ecosystems and then TWS consist of vulnerabilities in biological diversity, drying rivers, lakes, and wetlands, dust and sand storms, as well as other environmental dilemmas (Darpö 2021; Immovilli et al. 2022; Tanasescu 2022). Furthermore, this recognition causes ecosystems to be regarded as independent stakeholders with rights to existence, sustainability, healthiness, protection, and restoration—e.g., as shown in Hillebrecht and Berros (2017).

To sum up, although the ecosystem does not yet have legal rights, meeting its water needs is critical for sustainability. Emphasizing the perspectives of the ecosystem means providing for the environmental needs of water bodies such as rivers, wetlands, and groundwater, as well as including the stakeholders who live in the environment. In other words, reaffirming the ecosystem as a security referent object requires mainly nature-based policies, with actions aimed at providing for the needs of the environment and the diversity of species. Transboundary water allocation and use will thus not meet the needs of jurisdictions such as BCUs, instead it concentrates beyond artificial or ethno-territorial and socio-cultural borders. Although difficult to achieve, the sustainability and robustness of TWS where the ecosystem is considered the reference point is expected to be higher due to mitigating the environmental consequences of transboundary water management. For attaining sustainable TWS, in conclusion, all referent objects have to be considered simultaneously in designing policy pathways.

6.2.3 TWS for Whom? Toward Water Security in Transboundary Systems

By investigating some theories in International Relations and water security research, four TWS referent objects and their threat factors were explained in previous sections. These investigations provide insights for reframing TWS as a novel analytical framework that juxtaposes and synthesizes those pieces of theories. Hence, reframing TWS is based on the recognized security referent objects, threatening factors, and relevant actions for providing TWS (Table 6.1). The described investigations indicate differences among security referent objects in terms of sustainability, while security seems to be more robust and sustainable as it shifts from state to development, society, and ecosystem. Instead of focusing reductively on one of the referent objects (e.g., the ecosystem, which has higher sustainability), however, it clarifies that a comprehensive consideration of security referent objects is needed to ensure attaining sustainable TWS.

Examining the TWS referent objects in the developed framework indicates that the policies and actions depend on the chosen referent object, and the sustainability of the security provided increases gradually along a spectrum. However, due to the complex nature of water systems, the answer to the question "security for whom?" is "security for all" because in complex water systems, it cannot be expected that when one component is optimized, the whole system becomes secure. Therefore, to move toward TWS, all references must be considered simultaneously so that policies and actions are aimed at establishing security for the entire system. Therefore, policies and actions should not emphasize only one or two referent objects, since instability in transboundary basins is anticipative, despite the differences in the context of the relations.

Table 6.1 The foundations of TWS (created by the authors)

Security referent object	State	Development	Society		Ecosystem
Threatened factor(s)	Nation-State	Improving the agricultural and energy sectors, etc.	Identity	Ideology	Water Bodies
Most relevant action for providing security	Populistic measures, e.g. water nationalism	Water infrastructures	Moving toward solidarity of riparians	Alignment of the wants and the belongings	Providing ecosystem needs and rights
Level of security sustainability	Low ⟶ High				

6.2.4 TWS by Whom? The Providers of TWS

International Relations adopt a variety of perspectives on TWS providers. According to negative security, states are the central providers of security in transboundary basins, although their role attracts criticism. For example, Buzan (1991) argues that it will be difficult for the state to take part in multiple aspects of security and a variety of situations because its intervention may become a source of violence and insecurity. This is also reaffirmed in water resource management, where it is argued that top-down interventions by states neglect social needs, perspectives, and future vision (Jafari et al., forthcoming), and they may cause unexpected conflict-prone reactions (Fuchs et al. 2003; Oishi 1995). Despite the criticism directed toward the role of states, their participation in addressing security referent objects is still important. In other words, while states may be a source of violence, they are also obligated to other security referent objects. For instance, according to Kerr (2010), not all states are violent toward their own people, and many states are changing to accommodate security referent objects as part of their security agenda.

In contrast, the positive security perspective emphasizes other security providers, such as international institutions, NGOs, and civil society (Richmond 2002). Following the positive security discourse, water security research continuously seeks to be more integrated with social changes, acceptance, and needs in defining and providing water security (Gunda et al. 2019). As mentioned above, the need for development, societal dynamics, and the sustainability of the ecosystem are vital while mostly neglected when states are merely considered as TWS referent objects. Hence, a simultaneous consideration of all referent objects is necessary in order to provide TWS where states and other stakeholders play a crucial role. Considering all security referent objects and their contribution is thus indispensable to providing water security in transboundary basins. In order to be robust against socio-political and environmental changes, as well as consider all referent objects and their linkage to security, a "nexus security" approach is crucial (Beck and Villarroel 2013).

Nexus security is a holistic approach which focuses on both negative and positive security and the complexity of interactions between security referent objects for attaining sustainable security. Hence, the nexus approach emphasizes the need for a more holistic understanding of the dynamics between security referent objects and managing the complex interactions within and between systems at multiple levels, from domestic to regional and international scales. Moreover, security in transboundary basins is not provided by merely addressing the needs of either the state or those of other referent objects (e.g., development, society, or the ecosystem). When water policies come to nexus security, in contrast, the likelihood of attaining more sustainable TWS increases. In fact, there is a need to provide security in the whole system compared to fragmented reductionist policies, and to take a more comprehensive approach toward establishing and sustaining TWS. Thus, the nexus security concept has the ability to enhance cross-sectoral connectedness, reduce trade-offs, build synergies, improve governance, and protect ecosystems.

Following the notion of nexus security, TWS may be able to take the connectedness between security referent objects into considerations, which in practice means supporting the joint identification of synergies and actions that can reduce contentious water issues in transboundary basins. Hence, nexus security potentially enhances resource efficiency and good governance in transboundary basins (Strasser et al. 2016) while also addressing the criticism directed toward conventional frames of TWS by being more comprehensive. Moreover, the security nexus makes it possible to recognize threats to security referent objects basin-wide rather than focusing on domestic concerns in BCUs. The nexus security of transboundary basins may thus be more practical and more comprehensive in terms of sustaining shared-vision planning for security, as well as promoting cooperation in both intra-governmental and inter-governmental water relations.

6.3 Investigating Transboundary Water Security in Afghanistan

The theoretical investigation above indicates that understanding the foundations of water security and its overlooked components may be a gap that causes conflict and insecurity while co-riparians seek to provide TWS. This section applies the reframed TWS, using the influential components in security referent objects, threat factors, and measures for identifying the main referent object(s) in Afghanistan's transboundary waters. More specifically, this practical investigation seeks to identify the foundations of TWS in Afghanistan by examining security referent objects in order to identify the actions and policies required to provide water security in Afghanistan's transboundary basins, as well as pinpoint the consequences in different circumstances. Although Afghanistan is currently facing various political and economic security challenges, this does not prevent the country from analyzing the current situation and preparing for future challenges.

6.3.1 The Transboundary Water Situation in Afghanistan

Transboundary river basins cover more than 84% of Afghanistan's surface area, and at least 87% of the population of the country lives within one of these basins. Surface water flows out of Afghanistan at a rate of approximately 42.22 BCM per year (Malyar 2016). As a result, the utilization of water from rivers that originate in Afghanistan has a regional dimension (Favre and Monowar Kamal 2004). Figure 6.1 shows the river basins of Afghanistan.

This section describes the transboundary water arrangements of the area and analyses them with a broad focus on Afghanistan's transboundary water policies. The analysis determines which security referent object is referred to by each

Fig. 6.1 River Basin Map of Afghanistan (created by the authors)

Afghan policy and action. After recognizing them and using the nexus security, the sustainability of security in Afghanistan's transboundary waters is furtherly discussed.

6.3.2 *TWS for the State in Afghanistan*

Afghan governments view water infrastructure and dams as symbols of nation-state building and a way of staying in power (Baron 1975; Cullather 2002). The main reason for doing so is the multi-ethnicity of the country, which has never experienced a collective form of being together as a form of a single identity (Faraji-Rad et al. 2011). Nation-state building may thus be necessary for state security, prompting the Afghan government to use transboundary waters for this purpose. For instance, former President of Afghanistan Ashraf Ghani tried to use water to build a national identity to evoke a sense of patriotism and unity over water on a national scale (Mianabadi and Ghoreishi 2022). In other words, the Afghan transboundary water policy caused identity formation and inevitably resulted in identity otherization, mainly with Iran, impacting the Helmand and Harirod (Fig. 6.1) transboundary water relations (Ghoreishi et al., forthcoming).

Water-based otherization for unity and collective identity in Afghanistan, therefore, constructs "us" as Afghans and "others" as co-riparians, and it presents political opportunities for water nationalism and state-centric goals and security (Ghoreishi et al., forthcoming). Accordingly, these state-led policies and strategies have escalated contentious issues over shared waters and insecurity. However, constructing a collective identity from Afghanistan's Helmand and Harirod transboundary river basins entails erasing the previous collective identities of the entire basin and its inhabitants, such as the Greater Khorasan and the Greater Sistan, which have had a long history of coexistence. In this vein, destroying the previous collective identity has resulted in water conflicts between Iran and Afghanistan (Ghoreishi et al. 2021; Nagheeby and Warner 2022). In other words, Afghanistan's identity formation based on shared waters has positioned Iran as an enemy. Therefore, although the Afghan state served its short-term purposes during Ghani's presidency, Afghanistan's self-interested use of shared waters has caused political tensions between Iran and Afghanistan.

The roots of unsustainable water security in Helmand water interactions are fairly manifest because of that Afghanistan addressed the state as a main TWS referent object. For instance, it can be argued according to the Ghani speeches which were filled with nationalist rhetoric and populist slogans. At the 2017 conference on water and sustainable development at the Presidential Palace, Ghani claimed that "[transboundary] water is our dignity, and maintaining our dignity is our national goal" (Azadi Radio 2017). This state-centric water policy arguably follows the intensifying patriotism of the jurisdiction and fosters a "my waters" mentality in order to construct an otherized identity as a tool for unity. The construction of an otherized identity has intensified the ongoing water issues between the riparians. Thus, using waters for state-centric goals can impede TWS and basin solidarity.

The need for power is revealed in the point that focuses on the state as a security referent object. Afghanistan's government has actively sought to resume its hydraulic mission to gain more power in intergovernmental relations, especially from the late 1970s to 2021. Although other goals have been considered for the hydraulic mission as a silver bullet for Afghanistan's development, including food security, hydropower production, and mitigating the effects of droughts and floods (Thomas et al. 2016), the role of power in state-centric policies is prominent for the security of the state. As Loodin and Warner (2022) point out, Afghanistan's dam-building policies aim to secure bargaining power with co-riparians in transboundary basins. Therefore, protecting water, damming, and developing infrastructure has been crucial for Afghanistan to gain power (Nagheeby and Warner 2018). The impacts of these policies were destructive and conflict-ridden for co-riparians, and for instance, Afghanistan represented the releasing of shared waters toward Iran as a threat. Ghani claimed that "We will deal water in exchange for oil with Iran, releasing Helmand transboundary waters." By making such statements, the scope of decision-making was limited, and Iran and Afghanistan could not implement the 1973 Helmand Water Treaty. Thus, considering the state as the referent object of TWS in Afghanistan has created unsustainability for co-riparians and instigated conflicts in intergovernmental relations.

6.3.3 TWS for Development in Afghanistan

The development of transboundary waters in Afghanistan is often considered a TWS referent object. Since hydraulic missions in Afghanistan have served nation-building purposes (Baron 1975; Cullather 2002), the significant funds provided by the international community for water, energy, and food security have not brought security to the country. For instance, Afghanistan's development aid per capita was higher than that of any other country in the world during the 1960s. In 1955, Afghanistan was the first target of Premier Nikita Khrushchev's "economic offensive," the Soviet Union's first venture into foreign aid (Cullather 2002). Over USD 100 million in credit to Afghanistan financed a fleet of taxis and buses and paid for Soviet engineers to build airports, a cement factory, a mechanized bakery, a five-lane highway from the Soviet border to Kabul, as well as dams (Cullather 2002). Afghanistan had Soviet, Chinese, and West German dam construction companies and their financial support in this period (Cullather 2002). Furthermore, in the twenty-first century, Canada, Denmark, India, China, Turkey, and international organizations have all invested in the Helmand Valley Authority to build several large dams, diversion dams, drainage systems, irrigation canals, resettlement projects, and agricultural extensions (Ahlers et al. 2014). This financial support, the main objective of which was the establishment of security in Afghanistan, has failed to create a robust and sustainable situation in the different water-related sectors.

For instance, Afghanistan has the world's lowest electricity consumption and urgently needs to supply energy, with a consumption of 100 kWh per capita per year and only 30–38% of the population having access to the electricity grid (Ahmadzai and McKinna 2018). According to Afghanistan's power sector master plan, net demand is expected to rise from 2,800 GWh in 2012 to 15,909 GWh in 2032, with an average annual growth rate of 9.8% (World Bank 2018). Moreover, Afghanistan has suffered from the worst food insecurity since 2011 (FEWS NET Afghanistan 2019), with many households needing emergency assistance. However, the funds received were not used to ensure energy security, and the hydraulic mission has caused an increase in opium cultivation and production (UNODC 2018, 2021).

As previously stated, not satisfying desires and wants may jeopardize security. In this vein, hearing about obtaining funds would increase residents' wants, but because there will be no progress, they would become disappointed and lose hope. For instance, instead of dealing with energy and food security, Afghanistan's government inaugurated the Kamal-Khan diversion dam on the Helmand River in Nimruz province in 2021. Kamal-Khan is mainly a diversion dam that prevents water flowing toward Iran's border and the Hamoun wetlands, and it disrupts the natural waterway of the Helmand River. However, the irrigation canals of the Kamal-Khan dam have not been completed yet. No significant progress has been observed in completing the channels, and their construction work has either been halted or is very slow; it cannot be finished soon. People in Nimruz have difficulty obtaining water and are suffering from droughts and dust storms. In addition, the already completed sections are now under intense dune threat and are filled with sand. Thus, the Kamal Khan dam did not consider the development, societal, or ecosystem security referent objects and was inaugurated mainly to act as a state security referent object. According to Mianabadi et al. (2021), the Kamal Khan dam diversion will cause significant economic, human, and environmental harm for both downstream Iran and the inhabitants of the basin in Afghanistan.

To conclude, Afghanistan entered the twenty-first century with a government eager to focus on dam-building projects (Ahlers et al. 2014), and improving the water infrastructure was a strategic approach used to reflect national identity and patriotism. Consequently, the former Afghan state used water development to establish a collective national identity and strengthen its power over local authorities. Water nationalism, in other words, was set on Afghanistan's political agenda to deal with domestic issues, including providing security for the state, as well as development. Therefore, Afghanistan's government prioritized nation-state building by using transboundary waters over developing the country and meeting its developmental needs. Thus, societal and ecosystem security has been neglected in Afghanistan's transboundary water policies which impact the basin-wide TWS, inevitably and destructively.

6.4 Discussion

According to Table 6.2, Afghanistan's transboundary water policies for augmenting TWS are unsustainably framed. The practical investigations of some of Afghanistan's water policies and actions in its transboundary basins show that the state is the primary security referent object, with developmental issues considered at best. According to the theoretical foundations, ignoring the system as a whole and focusing on one of the referent objects rather than nexus security has been a contributing factor to water insecurity in Afghanistan. This argument is supported by the empirical evidence, according to which unilateral water exploitation and utilization, ad hoc policies of transboundary river development, and non-cooperative, empathetic relationships with co-riparians foster competitive attitudes and incompatibilities, resulting in protracted conflict-prone challenges to TWS.

Table 6.2 The foundations of TWS in Afghanistan (created by the authors)

Examples of Afghanistan's measures for providing TWS	TWS referent object			
	State	Development	Societal	Ecosystem
Kamal-Khan dam	×	×		
Irrigation channels		×	×	
International funds	×	×		
Preventing water flow toward the Hamoun wetlands	×			
Identity otherization	×			

TWS should be provided for all referent objects and stakeholders involved, based on the complex nature of water systems, while security is about the whole system (system of systems) rather than one referent object (e.g., state, development, society, or the ecosystem). The kind of provider needed for TWS is best promoted through participatory planning among co-riparians and collaboration. Such a nexus security approach could not effortlessly be established in Afghanistan so far, but it may be doable if nexus water security was considered. In conclusion, all references must be regarded as simultaneously and inevitably moving toward TWS in a nexus security manner, so that actions become aimed at establishing security for the entire system. Nexus security in transboundary waters causes policymakers to be aware of the effects of reframed components in providing water security, as well as to transcend artificial jurisdictions for broad considerations of security within natural boundaries. With this understanding, policy pathways for reframing TWS toward a sustainable, equitable, and peaceful way for both people and the ecosystem in Afghanistan are expected to be recognized.

6.5 Conclusion

The current understanding of TWS leads to policies that promote insecurity rather than security in transboundary waters, as well as pathways that facilitate the emergence of perpetual water conflict and disputes in riparian relations. Hence, to address contentious water-related issues in transboundary basins and policymaking to overcome insecurity, this chapter reframed TWS toward more comprehensiveness and suggested nexus security as the security provider. Reframing TWS partially transcends the conflict-generating contentions embedded in the ongoing understanding of water security in research and practice, and subsequently allows nexus security to be considered in policymaking for a more comprehensive action plan to establish security within transboundary basins and various security referent objects. Using the proposed framework to analyze transboundary water arrangements, neglected components could be revealed in on-ground analyses, and water policies may be further recognized for sustainable, equitable, and peaceful TWS for all security referent objects.

Applying the framework for assessing and scrutinizing nexus security in Afghanistan, this chapter identified the TWS referent objects in Afghanistan's transboundary water policies. Since Afghanistan's priority in the policy agenda has been nation-state building, the societal and ecological referent objects have been overlooked in establishing water security. According to a comprehensive understanding of nexus security, Afghanistan's water policies toward TWS have been unsustainable and have harmed other co-riparians. While hydraulic missions in Afghanistan, for instance, were supposed to provide security, the country's high insecurity, and its low water security index are related to a lack of a nexus approach to security. Overall, this chapter considers nexus security as indispensable to achieving TWS and recommends that policies for transboundary waters consider all security referent objects.

These theoretical and practical results guide the authors in providing suggestions for future research. With this approach to nexus security, the causes of perpetual conflicts and the insecurity of transboundary basins can be analyzed both at the subnational and international levels. This framework, as shown in the case study of Afghanistan, can examine the role of water nationalism, hydraulic missions, and populist rhetoric in water security. Applying this framework to other transboundary basins can further evaluate its applicability. For example, its application in basins such as the Nile, Aral, Volta, Euphrates-Tigris, Jordan, and other basins engaged in water conflict can identify the causes of insecurity by investigating TWS referent objects. Moreover, its application in the Great Lakes, Rhine, and Danube basins is likely to expand views on ecology as a security referent object and maybe empirically develop the notion of nexus security. Finally, in addition to these proposed case studies, the authors recommend global literature reviews to develop and evaluate the applicability of the framework as well as nexus security.

References

Ahlers R, Brandimarte L, Kleemans I, Sadat SH (2014) Ambitious development on fragile foundations: criticalities of current large dam construction in Afghanistan. Geoforum 54(1):49–58. https://doi.org/10.1016/j.geoforum.2014.03.004

Ahmadzai S, McKinna A (2018) Afghanistan electrical energy and trans-boundary water systems analyses: challenges and opportunities. Energy Rep 4(1):435–463. https://doi.org/10.1016/j.egyr.2018.06.003

Albrecht T, Varady R, Zuniga A et al (2018) Unraveling transboundary water security in the arid Americas. Water Int 43(8):1–39. https://doi.org/10.1080/02508060.2018.1541583

Allouche J (2005) Water nationalism: an explanation of the past and present conflicts in Central Asia, the Middle East, and the Indian sub-continent. Dissertation, University of Geneva

Allouche J (2019) State building, nation making and post-colonial hydropolitics in India and Israel: visible and hidden forms of violence at multiple scales. Polit Geogr 75:1–7. https://doi.org/10.1016/j.polgeo.2019.102051

Allouche J (2020) Nationalism, legitimacy and hegemony in transboundary water interactions. Water Altern 13(2):286–301

Anderson B (1983) Imagined communities: reflections on the origin and spread of nationalism. Verso, London

Asian Development Bank (2016) Asian Water Development Outlook 2016: strengthening water security in Asia and the Pacific. Philippines. https://www.adb.org/sites/default/files/publication/189411/awdo-2016.pdf. Accessed 26 Dec 2022

Azadi Radio (2017) Water is our dignity and maintaining our dignity is our national goal. https://da.azadiradio.com/a/28351411.html. Accessed 26 Dec 2022

Bakker K (2012) Water security: research challenges and opportunities. Science 337(6097):914–915. https://doi.org/10.1126/science.1226337

Bahaddin B, Mirchi A, Ahmad S et al (2018) System archetypes in water resource management. In: World Environmental and Water Resources Congress. https://doi.org/10.1061/9780784481400.012. Accessed 26 Dec 2022

Baron L (1975) The water supply constraint: an evaluation of irrigation projects and their role in the development of Afghanistan. Dissertation, McGill University
Beck MB, Villarroel WR (2013) Nexus security: governance, innovation and the resilient city. Front Environ Sci Eng 7(5):640–657. https://doi.org/10.1007/s11783-013-0549-5
Bigo D (2002) Security and immigration: toward a critique of the governmentality of unease. Alternatives 27(1):63–92. https://doi.org/10.1177/03043754020270S105
Bubandt N (2005) Vernacular security: the politics of feeling safe in global, national and local worlds. Secur Dialogue 36(3):275–296. https://doi.org/10.1177/0967010605057015
Burke A (2012) Security. In: Devetak R, Burke A, George J (eds) An introduction to International Relations. Cambridge University Press, Cambridge, pp 160–171
Buzan B (1991) People, state and fear: an agenda for International Security Studies in the post-Cold War era. Longman, Essex
Buzan B, Wæver O, De Wilde J (1998) Security: a new framework for analysis. Lynne Rienner, Boulder
Conker A (2018) Understanding Turkish water nationalism and its role in the historical hydraulic development of Turkey. J Nationalism Ethn ISSN 46(5):877–891. https://doi.org/10.1080/00905992.2018.1473353
Cullather N (2002) Damming Afghanistan: modernization in a buffer state. J Am Hist 89(2):512–537. https://doi.org/10.2307/3092171
Darpö J (2021) Can nature get it a right? A study on rights of nature in the European context, European Parliament. https://www.europarl.europa.eu/thinktank/en/document/IPOL_STU(2021)689328. Accessed 26 Dec 2022
Dinar S (2002) Water, security, conflict, and cooperation. SAIS Rev 22(2):229–253
Dinar S (2003) Preventive diplomacy, international relations, conflict resolution and international water law: implications for success and failure of the Israeli-Palestinian water conflict. Int J Global Environ Issues 3(2):188–225. https://doi.org/10.1504/IJGENVI.2003.003864
Eftekhari A (2013) Societal security: Islamic approach. Institute for Humanities and Cultural Studies, Tehran
Faraji-Rad A, Darkhor M, Sadati H (2011) Analyzing the process of nation-state in Afghanistan and its challenges. Sci Res J Manag Syst 4(2):113–133
Favre R, Monowar Kamal G (2004) Watershed Atlas of Afghanistan. http://www.cawater-info.net/afghanistan/pdf/afg_wat_atlas_part_1_2.pdf. Accessed 26 Dec 2022
FEWS NET Afghanistan (2019) Afghanistan Food Security Outlook. https://fews.net/sites/default/files/documents/reports/AFGHANISTAN_Food_Security_Outlook_10_2019.pdf. Accessed 26 Dec 2022
Fuchs C, Hofkirchner W, Klauninger B (2003) The dialectic of bottom-up and top-down emergence in social systems. https://papers.ssrn.com/sol3/papers.cfm?abstract_id=385300. Accessed 26 Dec 2022
Ganoulis J, Fried J (2018) Transboundary hydro-governance. Springer, Cham
Gharashi Y, Dehghani Firouzabadi J (2013) Theory and method in securitization studies. Strateg Stud Q 15(58):7–42
Ghoreishi SZ, Mianabadi H, Parvaresh-Rizi A (2021) The TWINS framework for illustrating co-existing conflict and cooperation in the Hirmand River Basin. Iranian J Soil Water Res 52(1):273–300
Ghoreishi SZ, Mianabadi H, Warner J, Nagheeby M, Vij S, Parvaresh Rizi A et al (forthcoming) Maintaining status quo or realizing transformation in transboundary water conflicts? Power-interests-identity nexus in the Helmand river basin
Gjorv G (2012) Security by any other name: negative security, positive security, and a multi-actor security approach. Rev Int Stud 38(4):835–859. https://doi.org/10.1017/S0260210511000751
Gleditsch N (2003) Environmental conflict: Neomalthusians vs. Cornucopians. In: Brauch H, Liotta P, Marquina A, Rogers P (eds) Security and environment in the Mediterranean. Springer, Berlin, pp 476–485

Grey D, Sadoff C (2007) Sink or swim? Water security for growth and development. Water Policy 9(6):545–571. https://doi.org/10.2166/wp.2007.021

Gunda T, Hess D, Hornberger GM, Worland S (2019) Water security in practice: the quantity-quality-society nexus. Water Secur 6(100022):1–6. https://doi.org/10.1016/j.wasec.2018.100022

Hanjra M, Qureshi M (2010) Global water crisis and future food security in an era of climate change. Food Policy 35(5):365–377. https://doi.org/10.1016/j.foodpol.2010.05.006

Hillebrecht A, Berros M (2017) Can nature have rights? Legal and political insights. https://www.environmentandsociety.org/perspectives/2017/6/can-nature-have-rights-legal-and-political-insights. Accessed 26 Dec 2022

Hussein H, Conker A, Grandi M, Hussein H (2020) Small is beautiful but not trendy: understanding the allure of big hydraulic works in the Euphrates-Tigris and Nile waterscapes. Mediterr Politics 27(3):297–320. https://doi.org/10.1080/13629395.2020.1799167

Huysmans J (2002) Defining social constructivism in Security Studies: the normative dilemma of writing security. Altern 27:41–62. https://doi.org/10.1177/03043754020270S104

Immovilli M et al (2022) Exploring contestation in rights of river approaches: comparing Colombia, India and New Zealand. Water Altern 15(3):574–591

Jafari M, Mianabadi H, Nasri Fakhredavood S, Mirhashemi Dehkordi SS, Ghorbani A (forthcoming) Constructive change by changeful means: an investigation into approaches to deal with water conflicts

Jervis R (1978) Cooperation under the security dilemma. World Politics 30(2):167–214. https://doi.org/10.2307/2009958

Kalpakian J (2000) Identity, conflict and cooperation in international river basins. Dissertation, Old Dominion University

Kerr P (2010) Human security. In: Collins A (ed) Contemporary security studies, 3rd edn. Oxford University Press, Oxford, pp 121–135

Larsen MAD, Drews M (2019) Water use in electricity generation for water-energy nexus analyses: The European case. Sci Total Environ 651(2):2044–2058. https://doi.org/10.1016/J.SCITOTENV.2018.10.045

Loodin N, Warner J (2022) A review of hydro-hegemonic dynamics on the transboundary Harirud River Basin: 2001–present. Water 14(21):1–17. https://doi.org/10.3390/w14213442

Malyar I (2016) Transboundary water institutions in developing countries: a case study in Afghanistan. MSc thesis, Oregon State University, Corvallis

McCracken M, Wolf A (2019) Updating the register of international river basins of the world. Int J Water Resour Dev 35(5):732–782. https://doi.org/10.1080/07900627.2019.1572497

McSweeney B (1984) Security, identity and interests. Cambridge University Press, Cambridge

McSweeney B (1998) Durkheim and the Copenhagen School: a response to Buzan and Waever. Rev Int Stud 24(1):137–140

Menga F (2016) Domestic and international dimensions of transboundary water politics. Water Altern 9(3):704–723

Mianabadi H, Alioghli S, Morid S (2021) Quantitative evaluation of "No-Harm" rule in international transboundary water law in the Helmand River Basin. J Hydrol 599:1–15. https://doi.org/10.1016/j.jhydrol.2021.126368

Mianabadi H, Ghoreishi SZ (2022) The role of water in otherization between Iran and Afghanistan. Institute for East Strategic Studies. https://www.iess.ir/en/analysis/3008. Accessed 26 Dec 2022

Mianabadi H, Ghoreishi SZ (2021) Reframing water security components. Iran Water Resour Res 17(1):239–261

Millennium Ecosystem Assessment (2005) Ecosystems and human wellbeing. https://www.millenniumassessment.org/documents/document.356.aspx.pdf. Accessed 26 Dec 2022

Molle F, Mollinga P, Wester P (2009) Hydraulic bureaucracies and the hydraulic mission: flows of water, flows of power. Water Altern 2(3):328–349

Nagheeby M, Warner J (2018) The geopolitical overlay of the hydropolitics of the Harirud. Int Environ Agreements: Politics, Law Econ 18(6):839–860. https://doi.org/10.1007/s10784-018-9418-9

Nagheeby M, Warner J (2022) The 150-year itch: Afghanistan-Iran hydropolitics over the Helmand/Hirmand river. Water Altern 15(3):551–573
Oishi M (1995) Conflict resolution and development: a case study of domestic development and related conflicts in Malaysia. Dissertation, University of Bradford, Bradford
Panić B (2009) Societal security—security and identity. Western Balkans Secur Observer-English Ed 13:29–38
Rai SP, Wolf A, Sharma N, Tiwari H (2017) Hydropolitics in transboundary water conflict and cooperation. In Sharma N (ed), River system analysis and management. Springer Singapore, Singapore, pp 353–368
Reimer LE et al (2015) Transformative change: an introduction to peace and conflict studies. Lexington Books, London
Richmond OP (2002) Maintaining order, making peace. Palgrave Macmillan, London
Roe P (2008) The "value" of positive security. Rev Int Stud 34(4):777–794. https://doi.org/10.1017/S0260210508008279
Sadoff C, Grey D, Borgomeo E (2020) Water security. In: Suggart HH (ed) Oxford research encyclopedias: environmental science. Oxford University Press, Oxford, pp 139–201
Sayar M (2012) The security and development nexus. Dissertation, College of New York
Srinivasan V, Konar M, Sivapalan M (2017) A dynamic framework for water security. Water Secur 1(1):12–20. https://doi.org/10.1016/j.wasec.2017.03.001
Strasser L et al (2016) Transboundary water resources in Afghanistan: climate change and land-use implications. Elsevier, Amsterdam
Šulović V (2010) Meaning of security and theory of securitization. Belgrade Centre for Security Policy. https://bezbednost.org/wp-content/uploads/2020/07/sulovic_2010_meaning_of_secu.pdf. Accessed 26 Dec 2022
Szalkai K (2012) Water issues are what states make of them: a constructivist approach to conflict and cooperation over trans-boundary waters. Dissertation, Central European University
Tanasescu M (2022) Understanding the rights of nature. The Deutsche Nationalbibliothek. Verlag, Bielefeld
Thapliyal S (2011) Water security or security of water? A conceptual analysis. India Q J Int Aff 67(1):19–35. https://doi.org/10.1177/097492841006700102
Thomas V, Azizi M, Behzad K (2016) Developing transboundary water resources: what perspectives for cooperation between Afghanistan, Iran and Pakistan. Afghanistan Research and Evaluation Unit Kabul, Afghanistan. https://www.researchgate.net/publication/303403362_Developing_transboundary_water_resources_What_perspectives_for_cooperation_between_Afghanistan_Iran_and_Pakistan. Accessed 26 Dec 2022
UN-Water (2013) Water security and the global water agenda: a UN-Water analytical brief. https://www.unwater.org/sites/default/files/app/uploads/2017/05/analytical_brief_oct2013_web.pdf. Accessed 26 Dec 2022
UNEP (2016) Transboundary river basins: status and trends. United Nations Environment Programme. http://www.geftwap.org/publications/river-basins-spm. Accessed 26 Dec 2022
UNGA (2022) The human right to a clean, healthy and sustainable environment. United Nations General Assembly https://documents-dds-ny.un.org/doc/UNDOC/GEN/G21/289/50/PDF/G2128950.pdf?OpenElement. Accessed 26 Dec 2022
UNODC (2018) Afghanistan Opium Survey 2017; challenges to sustainable development, peace and security. United Nations Office on Drugs and Crime. https://www.unodc.org/documents/crop-monitoring/Afghanistan/Afghanistan_opium_survey_2017_peace_security_web.pdf. Accessed 26 Dec 2022
UNODC (2021) Drug situation in Afghanistan 2021; Latest findings and emerging threats. United Nations Office on Drugs and Crime. https://www.unodc.org/documents/data-and-analysis/Afghanistan/Afghanistan_brief_Nov_2021.pdf. Accessed 26 Dec 2022
Waever O (1995) Securitization and desecuritization. In: Lipschutz RD (ed) On security. Columbia University Press, New York, pp 46–87

Waever O (2008) The changing agenda of societal security. In: Günter H et al (eds) Globalization and environmental challenges. Springer Berlin, Heidelberg and New York, pp. 581–593
Waever O, Buzan B, Kelstrup M, Lemaitre P (1993) Identity, migration and the new security agenda in Europe. London, Pinter Publishers
Wendt A (1999) Social theory of international politics. Cambridge University Press, New York
Williams JM (2018) Stagnant rivers: transboundary water security in South and Southeast Asia. Water 10(12):1819. https://doi.org/10.3390/w10121819
Wolf AT (1995) International water dispute resolution: the Middle East multilateral working group on water resources. Water Int 20(3):141–150. https://doi.org/10.1080/02508069508686465
World Bank (2018) Afghanistan renewable energy development issues and options. https://documents.worldbank.org/en/publication/documents-reports/documentdetail/352991530527393098/afghanistan-renewable-energy-development-issues-and-options. Accessed 26 Dec 2022
Zedner L (2003) The concept of security: an agenda for comparative analysis. Leg Stud 23(1):153–176
Zeitoun M et al (2016) Reductionist and integrative research approaches to complex water security policy challenges. Glob Environ Chang 39:143–154. https://doi.org/10.1016/j.gloenvcha.2016.04.010

Chapter 7
Water for Peace Revisited: Reconsidering the Role of Securitization in Water Cooperation

Ali Oguz Dirioz

Abstract This chapter identifies the elements that create a political and social environment conducive to water cooperation, using the case study of the Water for Peace (WfP) initiative by Türkiye and Israel in the early 2000s. The chapter is based on a historical analysis of the evolution of bilateral relations between the two countries, as well as secondary climate and water data to support the analysis. Water for Peace was an unsuccessful initiative that aimed to start collaboration on water between Türkiye and Israel, resting on the idea of using water for the de-escalation of tensions in the region by supplying water from Türkiye. This case shows how failing to involve civil society and particularly the business community undermines state-level efforts to tackle regional water-related challenges.

Keywords Water security · Water cooperation · Securitization · Türkiye · Israel

7.1 Introduction

This chapter analyzes the attempted cooperation on water between Türkiye[1] and Israel in the early 2000s with the help of various theoretical lenses. The chapter identifies what elements were missing for the cooperation to work and examines how these elements should be utilized if the cooperation were to be revived. A historical analysis of the evolution of Turkish-Israeli relations, as well as secondary climate and water data is used to support the arguments presented in the chapter. The two main conceptual perspectives applied are the regional attributes of water politics in

[1] Since 2022, the country is called the Republic of Türkiye in English, and no longer the Republic of Turkey.

A. O. Dirioz (✉)
International Entrepreneurship Department, TOBB University of Economics and Technology, Ankara, Turkey
e-mail: adirioz@etu.edu.tr

K. Szálkai and M. Szalai (eds.), *Theorizing Transboundary Waters in International Relations*, Springer Water, https://doi.org/10.1007/978-3-031-43376-4_7

the Middle East and the overall conceptual frameworks provided by the literature on securitization and de-securitization.[2]

The chapter focuses on the Water for Peace (WfP) initiative that was drafted in the 1980s and early 1990s as a collaborative undertaking in the Middle East through the use of water as an item of regional trade. This initiative was meant to sell water originating in southern Türkiye, to be transported via the Mediterranean Sea. Early discussions for this undertaking can be traced to the late 1980s, in the form of Ozal's 1986 Peace Water Pipeline (Gruen 2007). A water treatment facility that could serve this purpose was built in Türkiye by 1997. The project was officially initiated by signing an intergovernmental agreement at the level of foreign ministry senior officials in 2004 (Turkish Ministry of Foreign Affairs 2004), which envisaged transporting water from the Manavgat River to Israel by sea. However, the project was eventually shelved.

WfP should be understood as a significant attempted cooperation initiative in the Middle East. This project essentially had the potential to become successful, and the conditions to be replicated, modified, or re-initiated may still make it relevant. This chapter draws attention to the importance of maintaining dialogue and suggests possible actions for appeasing the tensions related to water security in the region. An important conclusion is that civil society can play a significant role in sustaining dialogue and that sustained dialogue may help reach a certain degree of understanding as well as possible temporary agreements, which may eventually be milestones in the course of adopting a more permanent strategy.

7.2 Cooperation on Tackling Water Scarcity in the Middle East and North Africa

7.2.1 Conceptual Framework

Water is a limited resource, and it is becoming even more scarce through the effects of climate change especially in regions such as the Eastern Mediterranean (East Med) and the Middle East and North Africa (MENA) region. Water, which is incidentally one of the seventeen Sustainable Development Goals (SDGs) of the United Nations, is a basic resource needed for life. It plays a vital role in life and agriculture, and it is needed to sustain the overall well-being of society. For these reasons, water security is important, especially in regions where it is difficult to access water resources.

Water security concepts are at the intersection of multiple issues, such as the environment, agriculture and food, public health, as well as energy security (Kibaroğlu et al. 2007). The terminology used by UN-Water provides a general working definition of water security:

---

[2] This chapter was drafted before the 6 February 2023 earthquakes in Türkiye. The earthquakes would further reinforce the conditions argued in this chapter, i.e. the need for more water by Turkiye and cooperation on development and environmental issues.

> The capacity of a population to safeguard sustainable access to adequate quantities of acceptable quality water for sustaining livelihoods, human well-being, and socio-economic development, for ensuring protection against water-borne pollution and water-related disasters, and for preserving ecosystems in a climate of peace and political stability. (UN-Water 2013)

Water security can be defined in a variety of ways, by: (a) using an interdisciplinary approach that emphasizes the sustainability of an entire water basin's environment (as discussed in Cook and Bakker 2012) from an anthropocentric standpoint, (b) concentrating on the requirements of society, such as the availability of water for households, agriculture, and industry (Kibaroğlu et al. 2007), and (c) adopting a more integrative principle, highlighting the availability of clean, safe, potable, and sanitary water (Sikri 2010).

In this context, beside defining water security, it is useful to conduct a conceptual consideration of "securitization" and "de-securitization". Securitization can be understood as discourses and processes that construct and frame an issue (e.g. water scarcity) as a matter of security, thereby moving that issue outside the range of usual political action (Buzan et al. 1998). The securitization of an issue is hence the construction of a particular policy item or reference as an existential threat or national security challenge and is presented as such to the intended audience, which in turn can legitimize enacting emergency measures. De-securitization can be seen as the reverse process, i.e. moving an issue "away from exceptional and back to normal spheres of politics" (Aggestam 2015; Frölich 2020).

Securitization can be a barrier to cooperation (e.g. through the internal politicization of water issues regarded as important for national security and seen as an element of power or resource nationalism), or it can be a driver of change though mobilizing governments and societies into re-assessing perceived threats to the environment. In this respect, civil society and the business community have a great role to play. As the elements directly affected by issues related to water (e.g. the quantity and/or quality of the water available for use by the population and the industry), civil society and the business community can have first-hand knowledge on how to tackle water issues and create public pressure on governments to act and incorporate public demands. As such, the feedback of non-governmental actors and their participation in the formulation of water-related policies is important (Diriöz 2020).

As discussed earlier, securitization erects barriers against cooperation that tackles regional issues. Civil society can assertively remove such barriers by putting pressure on politicians to increase mutual trust and understanding, or work through international bodies, enterprises, and other civil society elements to attain a common purpose through various arrangements. Conversely, civil society can use the innate sense of urgency and threat perception of securitization to mobilize governments by framing environmental and water issues as security concerns that need to be addressed immediately.

7.2.2 *The History of Water Cooperation in the Middle East*

The Middle East is a highly water-stressed region, and access to adequate water supply is viewed as a strategic asset, which leads to the securitization of water resources. This securitization, combined with a fragile regional atmosphere characterized by chronic political instability and armed conflict that breeds entrenched distrust between various actors, diminishes the chances of cooperation (Keohane and Nye 2001). In such an atmosphere, sustained dialogue is a starting point and a minimum pre-condition for any hope of reaching eventual cooperation.

Access to water projects are areas where a combination of private and public partners, as well as international investors and organizations, civil society, citizens, and businesses can play a role. This is another element of focus in the region, and the revisited collaboration between Türkiye and Israel on water issues could also yield results.

Consequently, there have been few attempts at water cooperation, which makes it possible for water to be securitized as a vital resource. Türkiye and Israel, the actors who tried to cooperate through the WfP initiative, were involved in most of these cooperation attempts, either through their government, or through their civil societies and businesses. Some of the other regional cooperation attempts are described in the following sections.

7.2.2.1 Study on Integrated Development in the Jordan Rift Valley (JRV)

As part of the 1994 Peace Agreement between Jordan and Israel, the Jordan Valley Authority (JVA) was established and given responsibility for the development and management of water resources in the Jordan Rift Valley, including functions such as irrigation, dam management, wastewater reuse development, soil conservation, the protection of the natural resources of the Dead Sea, as well as the management and protection of land in the Araba Valley for touristic, industrial, and agricultural purposes (Mumssen and Triche 2017). The study the JVA conducted was a cornerstone of the 1994 Peace Agreement and a small yet significant form of collaboration that took place between the two nations. It is also significant because it laid the foundation for the need for a conveyor to feed the Dead Sea. Similarly to WfP, it also rests on the idea of 'feeding' water from another source to save a body of water.

One important conclusion of the JRV study was the need to find a solution for the over-use of the water flowing into the Dead Sea, indicating a need for the integrated development and management of water and the environment in the Jordan Rift Valley. This led to an attempt to have conveyors feed water to the Dead Sea, in the form of the Red Sea-Dead Sea Conveyor (RSDSC).

7.2.2.2 The Red Sea-Dead Sea Conveyor (RSDSC)

The Red Sea-Dead Sea Conveyor (RSDSC) is the most well-known case of water cooperation in the Middle East. A pre-feasibility study was conducted in 2005 (Willner et al. 2013), and the project was to be initiated by Jordan, together with Israel and the Palestinian Authority, following an agreement between Jordan and Israel in 2013 (TOI 2021). However, the project has been on hold since then.

The project had three objectives: (1) save the Dead Sea from environmental degradation, (2) desalinate water and generate energy, and (3) build a symbol of peace and cooperation in the Middle East (Allan et al. 2014; Glausiusz 2013). The project envisaged transporting water from the Red Sea to stop the shrinkage of the Dead Sea, which would create new territorial issues by creating new land.

The RSDSC is an example of how an environmental issue can be perceived as an immediate challenge. According to the water security concept of UN-Water (2013) and Kibaroğlu et al. (2007), the ability to fulfil societal needs such as agriculture, food, water for sanitation, as well as energy is under pressure from climate change and environmental issues. In the context of the environment, securitization can be a barrier to cooperation through internal politicization and resource nationalism. By contrast, it could be logically suggested that through the securitization of the environmental challenge itself, actors may be induced to collaborate.

The debate continues whether cooperation can occur despite securitization or through the securitization of the environmental challenge, which can lead to collaboration among the actors involved. In either case, cooperation can be very difficult to achieve, especially in such grand projects. By contrast, smaller-scale, limited cooperation on specific projects such as the Industry for Peace initiative, creating the Turkish Model Industrial Zone in the West Bank, or infrastructure projects for water can perhaps be objectives that can be achieved.

Access to clean water, water treatment, and sanitation projects are areas where a combination of private and public partners, as well as international investors and organizations, can potentially play a greater role. This is another element of focus in the region, and a revisited collaboration between Türkiye and Israel on water issues could yield results that are similar to those of the As Samra project.

7.2.2.3 As Samra

The expansion of the As Samra wastewater treatment plant was started by the Jordanian government in 2012, under the direction of the Ministry of Water and Irrigation (MWI). International funders including the Millennium Challenge Corporation and private lenders such as a regional consortium run by the Arab Bank provide funding for the project. Samra Wastewater Treatment Plant Company Ltd. (SPC) and Samra Plant Operation and Maintenance Co. Ltd. (O&M) are the two project businesses, in the form of joint ventures between Morganti, and Suez Environment, and its subsidiary, Degrémont. International awards have been given to the creative finance package, which was created to reduce risks. As Samra is the first project of its kind in the Middle East to combine sponsor, host government, and private funding (Mumssen and Triche 2017).

7.2.2.4 The Disi-Amman Conveyor

Another grand project that was established in the region through a multinational consortium that involved a Turkish company was the water conveyor from the Disi aquafer in the Jordan-Saudi Arabia border area to Jordan (often referred as the Disi-Amman Conveyor or simply the Disi Conveyor).

The Disi-Amman Conveyor is an international build-operate-transfer (BOT) project to construct and run a 325-km pipeline to transport water from a non-renewable aquifer beneath the Disi region near the Saudi border (Wadi Rum) to Amman, Aqaba, and more recently, Zarqa. For a total of 25 years, it will transport 100 million m^3 of water every year. The Disi Water Company (DIWACO), which is owned by GAMA Energy Water International B.V. (50 percent by GAMA Holding, a Turkish Company, and 50 percent by a subsidiary of General Electric), was in charge of the majority of the construction phase. Disi Amman Operation & Maintenance (DAOM), a branch of Suez Environment, operates the French Disi-Amman Conveyor company. The initiative became operational in 2013, and it became commercial in 2014 (Mumssen and Triche 2017). Disi was thought to be a significant supply of water that could be supplied to the Amman capital region in the north, although it would not satisfy either household consumption or agriculture under the current influx of refugees.

7.2.2.5 Good Water Neighbors in the Kidron/Nar Valley Project

In contrast with the above grand projects, which were mostly within the sovereign territory of one country, the needs of local communities often require cooperation on a smaller scale that nevertheless surpasses the boundaries of any specific country. The contents of the Good Water Neighbors (GWN) project were determined by exactly such needs.

EcoPeace Middle East established Good Water Neighbors (GWN) in 2001 to elevate cognizance of the shared water needs of Jordanians, Palestinians, and Israelis. The project develops political will for transboundary cooperation on water and sanitation. The technique of the GWN is based on an authentic concept that primarily relies on enticing cross-border groups and using their interdependence on shared water sources to increase communication and cooperation for sustainable water control and to advance peacebuilding (Djernaes 2020). GWN is not a particularly big project, but it is relevant as a form of cooperation initiated by civil society.

These examples demonstrate how Türkiye and Israel, either as a state or through their public and businesses, have been involved in regional water-related cooperation, and that they can therefore be considered the actors in the center of the water cooperation debates. Additionally, these examples further accentuate the potentially constructive and useful role of international organizations, governments, public–private partnerships, investors, non-governmental organizations, and the general public (i.e. water users) in water cooperation. This indicates that effective regional water cooperation requires the adoption of a holistic approach that considers the needs and goals of each actor at the international, national, and local level, and what they can bring to the table regarding specific water cooperation projects.

7.3 Discussion: The History, Failures, and Possible 'Evolved' Revival of WfP?

Regarding the securitization of water, the relations between Türkiye and Israel can be considered an example for the Water for Peace (WfP) initiative due to the agreements made over water in the past, and their location in a water-scarce Eastern Mediterranean and Middle East. In such water-scarce regions, water has been identified as a highly securitized vital environmental resource, prone to conflicts and contestations (Pradhan & Srinivasan 2022), which can also be observed in nearby regions, such as the Fergana Valley in Central Asia (Pradhan & Srinivasan 2022). Since a history of attempted cooperation has been identified in the region that involves Türkiye and Israel as the central actors, the lessons learned from past experiences suggest that even a highly securitized resource such as water can still potentially yield future cooperation, even in complex political and geographic locations such as the Middle East.

During the 1980s and early 1990s, a project was envisaged to combine forces regarding water issues between Türkiye and Israel. The idea rested on the de-escalation of tensions between Israel and its neighboring countries by selling water from Türkiye, thus reducing the issues connected to transboundary waters (Gruen 2007). The project was often referred to as the Water for Peace initiative and envisaged selling water from the Manavgat River in Türkiye to Israel. Although this project has not been implemented, significant water treatment facilities have been built.

Time has revealed that trade between Türkiye and Israel has been developing despite the political tensions in Turkish-Israeli relations since 2009. Bilateral trade has grown resiliently, with dynamic relationships sustained by the business sector over the last decade. From 2020 to 2021, the volume of trade increased from USD 6.2 billion in 2020 to USD 8.4 billion in 2021, making Israel the eleventh most important destination for Turkish exports (Daily Sabah 2022; TUIK 2022). Within this framework, it would not be wrong to say that a water and agriculture policy could be developed in a similar way, separately from day-to-day politics because these areas entail financial gains for both sides (Diriöz 2020). In addition, WfP could be formulated in a way that would change the quality of politics as well (and thus prevent politics from sabotaging economic cooperation) by envisioning more room for involving support from civil society and taking advantage of the consistent Turkish-Israeli economic ties in the context of the private sector.

Communication must continue for a revival of WfP, and both civil society and the business community should play a greater role. However, in a revised plan, the option of transporting significant quantities of water may no longer be possible due to the severe water shortages and the alarming environmental conditions that prevail today. Rather, a symbolic message can be conveyed by transporting a modest quantity. The need for communication was expressed in the March 2022 meeting between the Turkish and the Israeli President (Kirişci and Arbell 2022), when both parties highlighted their desire for co-existence despite the historical burden between the two countries. However a heavy load it may be, both countries prefer to start a campaign of honor and reliance for the greater good, as well as deepening cooperation in many fields (Israel Ministry of Foreign Affairs 2022). The Turkish Foreign Minister's visit to Israel in May 2022 is also considered a historic turning point (Anadolu Ajansı 2022), as it was the first ministerial-level visit from Türkiye to Israel for over a decade and thus a clear sign that relations were on a course of normalizing. Re-establishing ambassadorial-level relations was also key. Therefore, although grand projects may still be difficult to achieve, there may be further cooperation. Turkish-Israeli relations have not overcome all the existing problems with the two visits, but the potential trend to collaborate will have a positive impact in a securitized region.

The WfP initiative implicitly assumed in its initial stages that the water resources available from Türkiye would be sufficient to provide adequate levels of water for both Palestine and Israel. Current projections suggest that this may no longer be the case (Zittis et al. 2022). For example, the growing number of sinkholes appearing on the Konya Plain in Türkiye, and particularly the Karapinar basin (Dursun 2022), suggest that the nature of the cooperation may be increasingly relevant for technologies on agriculture and effective water management. The effects of the involvement of civil society in matters of such importance should never be underestimated. These organizations could use their resources to put pressure on their governments via influencing society, or even international organizations, so that the desired results can be achieved regarding natural resources.

Considering the challenges of climate change discussed above, the decrease in the amount of available water puts stress on water for consumption by households, the industry, and agricultural irrigation. The Eastern Mediterranean and the Middle East regions are already among the most water-scarce regions. Water-sector investments in the Middle East are often challenging issues, complicated by transboundary waters and unequal water distribution, while climate change further accentuates inequality both within nations and among nations (Badran 2017; Sümer 2014). Although it is relatively better placed in terms of access to water resources than the rest of the Middle East (von Lossow and Shattat 2020), Türkiye is also facing the problem of having less water per capita available annually due to a growing population. According to FAO's Aquastat, Türkiye's per capita renewable water resources fell from an average of 3,044 m^3 in the period 2008–2012 to 2,811 m^3 in 2013–2017 (FAO AQUASTAT 2020).

SUEN's recently published report (Eşbah 2021a) illustrates the need for better urban water conservation to promote more efficient use of water. The growing population of Turkish cities and the need for more agriculture puts water in increasingly

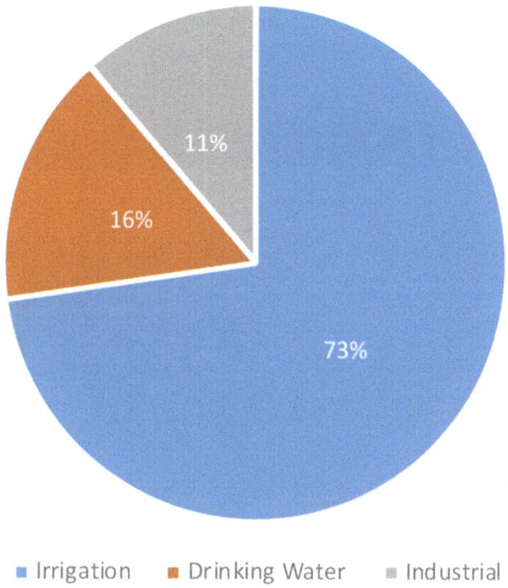

Fig. 7.1 Water consumption according to sectors in 2012 (SYGM 2016)

higher stress levels in meeting the needs of the country's population. Awareness still needs to be raised on water usage, as pre-rinsing dishes before using a dishwasher, for example, causes a waste of 57 L of water for each cycle (Finish Water Index 2021). The flash floods that devastate urban settings and infrastructure development, which require both investment as well as know-how, are other important fields for renewed collaboration across the region. Türkiye is an upstream state in the Middle East but a downstream state in the Balkans and faces flood in those regions. After earthquake-related deaths, floods are among the major challenges the urban population living in Turkish cities is facing, causing the second highest number of natural disaster-related deaths in Türkiye (Himat et al. 2019) (Figs. 7.1 and 7.2).

Especially in the face of various natural disasters (e.g. forest fires, mucilage, floods), the deterioration of the natural balance, and the growth of the population, the pressure in the agricultural sector in Türkiye is increasing day by day, and the importance of water resources is increasing as a result. Thus, the alarming environmental situation itself is also becoming a security issue. Similar situations are observed or expected in many countries. However, governments can take precautions against the severity of natural events *despite* securitization, or they can be motivated to take precautions *due to* the awareness created by securitization.

Israel, like Türkiye, is also facing environmental challenges and criticism. For many years, it was thought that the water resources available from Türkiye would be sufficient to provide adequate levels of water for both Palestine and Israel.

Until the mid-2000s, the stance of Turkish foreign policy towards the Middle East mostly comprised non-involvement, which allowed many to view Türkiye as a broker or intermediary in the Middle East peace process. The degradation of Turkish-Israeli relations over the last decade has been one of the reasons why the water for peace initiative has seemed a distant prospect.

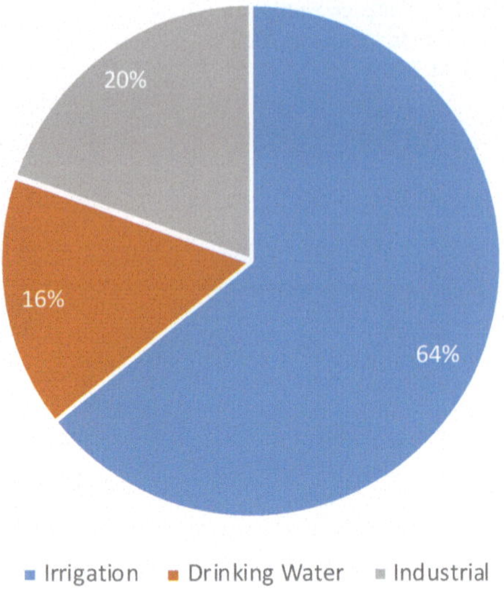

Fig. 7.2 Projected consumption in 2023 (SYGM 2016)

Türkiye is suffering from severe water shortages. This major change since the initiation of WfP, on top of the failure to realize the WfP project over the years, may make many cautious about calling for any revival of such initiatives. However, WfP, as well as the water treatment facility built on the Manavgat River as part of this project, has been one of the few examples of cooperation between these countries. In contrast, the realization of the industry for peace initiative on Jenin in the West Bank[3] has been a rare success despite the tense situation not only between Türkiye and Israel but also in the region overall.

According to Euromesco Joint Policy Study No. 15 on water security in the Middle East (Maghen and Kronich 2020), as well as the World Bank's statistics on water sector regulation in the Middle East and North Africa (MENA), these are some of the rare schemes of cooperation by nation states in the Middle East. As identified earlier, the absence of a strong role by civil society on water-related cooperation is striking. The presence of a Turkish-type industrial zone in the industry for peace initiative by the Union of Chambers and Commodity Exchanges of Türkiye (TOBB) gives an idea how civil society can make genuine efforts to enhance the cooperation between societies despite the political difficulties. Türkiye's and Israel's specific attempts at water-related cooperation over the years suggest that such cooperation may still be possible, and it is important to notice that the dynamics have changed over the years.

[3] The active and ongoing 'Industry for Peace' (*Barış İçin Sanayi*) initiative is run by The Union of Chambers and Commodity Exchanges of Türkiye (*Türkiye Odalar ve Borsalar Birliği—TOBB*). TOBB initiated the Industry for Peace plan (TOBB-BIS) to develop a Turkish-type industrial zone in Jenin, West Bank. TOBB-BIS was contracted in early 2010 for developing, managing and operating the Jenin Industrial Free Zone (JIFZ). TOBB-BIS established infrastructure and industrial facilities. The purpose is to foster trade and manufacturing between Türkiye, Palestine and Israel, thus promoting peace and stability. See the Palestinian Investment Promotion and Industrial Estates Agency: http://www.pipa.ps/page.php?id=28878fy2656143Y28878f.

In light of the changing water stress situation in Türkiye, and the ongoing pandemic of Covid-19 (which has increased hygiene-related water consumption), a 'revisited' water for peace initiative will be fundamentally different compared to when the initiative was originally envisaged, considering the different needs in Türkiye's water consumption. Unlike an industrial zone, which is essentially a construction and management project in a special zone, the water for peace initiative envisaged an exchange of potable water from freshwater resources between two countries that are not adjacent but are developing technological solutions for how they can exchange water as a symbol as well as an actual commodity.

The dystopian essay Brave New World Revisited by Aldous Huxley (1958) has in part inspired the title of this chapter. In many ways, especially from an eco-centric point of view, we are rapidly degrading the resources of our planet toward a dystopian future. In that respect, Water for Peace is also in dire need to be revisited due to the environmental urgency. Water-related cooperation to further the stability and prospects of peace in the region of Türkiye, Israel, and the Palestinian Authority should not only remain an option, it should become a necessity. As part of the cooperation, the exchange of water as a commodity may be of a lower quantity than previously hoped for. However, despite the lower quantity, its symbolic nature would denote the contribution of civil societies and business communities to lasting peace and stability, where Israelis and Palestinians, as well as the peoples of other regional countries such as the Turks and the Jordanians may be collaborating on water-related topics.

Beside the actual scheme for the cooperation, other dynamics will find the revisited water for peace initiative useful as a form of cooperation. First, it will still be a symbolic contribution for Türkiye to help the normalization of Palestinian and Israeli relations. Second, with the frequent changes of government in Israel and the nearly decade-long tense relations between Türkiye and Israel, there have been signs since 2021 of the will to at least have some form of cordial relations.

Until 2021, the supposedly tense relations between Türkiye and Israel were often intensified by the fossil fuel explorations on the Eastern Mediterranean. One main source of tension has been the East Mediterranean pipeline, which was supported by Israel and Greece as well as Cyprus (Greek Cypriot) and Egypt, which are parties to an Eastern Mediterranean Gas Forum. This forum, and the plans to have a pipeline over areas Türkiye considers its own territorial waters, has increased tensions in the region. Furthermore, not including Türkiye in the Eastern Mediterranean Gas Forum and dismissing Türkiye's continental shelf rights through Exclusive Economic Zones (EEZ or EZ) has isolated Türkiye in the regional cooperation on energy issues.

The (Greek) Cypriot administration's search for hydrocarbons in the claimed EZs, which are not recognized and strongly objected by Türkiye. However, tensions in the Eastern Mediterranean are not inevitable, and promoting trade and cooperation may possibly lead to an easing of tensions (Diriöz 2022). Cooperation with Türkiye could be possible with the help of new technologies on water and the environment, which are much needed to address the growing stress on water resources. Such cooperation on environmental and water-related issues would be a sign of the amelioration of the previously tense relations regarding energy. Israel is often proud to be a leader in new technologies on sustainable energy as well as new solutions for urban sustainability, and water consumption could be a new way in which the previous symbolic attempt to cooperate on water-related issues could evolve.

Considering increasing urbanization and climate change, the sustainable planning, design, and management of urban rivers has become an important issue all over the world. This issue is likely to remain relevant as urbanization and climate change continue and the need to improve sustainability and quality of life increases in the future. Climate change, which is accompanied by trends such as population growth, migration, rapid urbanization, and water scarcity, increases the tension and instability in the Mediterranean and the Middle East (Cahan 2017). Türkiye is one of the countries that stand to suffer the most from the inevitable effects of climate change and its ecological, economic, social, and political dimensions (Kadıoğlu 2019). Therefore, there is a need to develop adaptive strategies to overcome pressures on critical urban landscapes, particularly urban river corridors (Eşbah 2021a, b).

The cooperation attempts between Türkiye and Israel, as well as those between Israel and other countries of the region in the last ten years are worth considering, despite the relatively tense relations between these countries. Additionally, how the business community has gained momentum especially between Israel and Türkiye in terms of trade and collaboration is particularly promising, regardless of the constantly degrading political rhetoric. Therefore, based on the example of the industry for peace, cooperation on water could be possible, although such cooperation is not guaranteed to happen.[4]

7.4 Conclusion

This chapter has explored the conceptual framework of the role of securitization and water-related cooperation and diplomacy, in particular, how water and environmental cooperation could play a role in "watering down" regional tensions (Diriöz 2020). Regarding securitization and the notion of water security, a similar structure was used to that of Diriöz (2020). The chapter suggested that on the one hand, civil society and businesses were the missing component in the past, and they could be useful in revitalizing the WfP initiative. Furthermore, the envisaged cooperation rests on increased availability of water, and the current revitalized cooperation should make greater use of civil society and businesses in terms of cooperation on technology and investments.

The visit of the President of Israel to Türkiye and the Turkish Foreign Minister's visit to Israel, as well as re-establishing ambassadorial-level bilateral relations are

[4] Many Middle Eastern governments are wary of the activities of nongovernmental organizations (NGOs), sometimes considering them a threat to national security, especially if NGOs receive foreign funding. International organizations may increase the reputation of NGOs, although in some cases the effect could also be the opposite.

positive steps towards a sustained improvement of relations between the two countries. However, the pressure to restore cordial relations has been bottom-up and has come especially from the business community. Therefore, the business community, the scientific community, academia, and NGOs, all of which constitute civil society, can play a crucial role in the 'revisited' water for peace initiative. The implementation of integrated approaches such as the Water Energy Food (WEF) nexus approaches,[5] and cooperation based on technology and investment require support from the private sector and civil society. Earlier decades had envisaged Water for Peace as a solution to reducing regional political tensions and de-escalating water securitization through the provision of water by Türkiye (Gruen 2007). The current concept of a 'Revisited Water for Peace' places further emphasis on Israel's contribution. Mutual benefits could be attained by further including civil society and businesses by fostering sustainable development, economic growth, as well as overall regional cooperation to mitigate the effects of climate change. Such regional environment-centered mutual benefits, which are accentuated by the prospect of economic development, can make this revisited initiative more sustainable both regarding its durability, as well as its impact on the environment.

There is an existing motivation for cooperation at the state level between Türkiye and Israel on energy matters and energy trade (Dağ and Rezzakoğlu 2022), thus a symbolic water infrastructure could also be pursued along the same route (Türkiye has already built water infrastructure in the Turkish-controlled northern part of Cyprus[6]). However, the actual core area of cooperation could center on the WEF nexus approach projects and the modernization of urban settings (Eşbah 2021b). Furthermore, as the ongoing Russia-Ukraine war has demonstrated, due to the global vulnerability of food supply chains, cooperation on food, and especially more urban agriculture and other technologies are essential in the revisited notion.

Therefore, water for peace cooperation may still be revived, but the type of cooperation needs to be reconsidered to include civil society and private businesses in the transition to a circular or green economy. Additional factors need to be considered that simply did not exist back when WfP was first meant to be implemented, such as the Covid-19 pandemic, which has brought increased water consumption due to hygiene concerns. Other factors include climate change-related draughts and floods, tensions over the extraction of natural gas in the Eastern Mediterranean, and the frequently changing governments in Israel.

A pioneering project addressing both water and energy infrastructure would have symbolic value despite the potential challenges. However, sustained cooperation should be encouraged at the level of the civil society and private businesses so that the project is sustainable both in terms of bilateral relations and in terms of the

[5] The Water-Energy-Food (WEF) nexus approach is a method manage the same water resources for both the water consumption of the population for their domestic and industrial use, as well as for electricity generation and for irrigation in agriculture. WEF nexus approach intersects these sectors and requires a holistic understanding of the environment and how to use water resources in a sustainable way for electricity generation, agriculture, industrial and household usage. For more information, please look at Diriöz (2021).

[6] Türkiye recognizes the Turkish Republic of Northern Cyprus.

environment. Civil society and private businesses can play a critical role in Track-II and back-channel diplomacy due to their impact on governments and society. Despite the uncertainty regarding the actual water trade, water-related cooperation driven by civil society and businesses can remain relevant for the relations between Türkiye and Israel. Therefore, even a highly securitized vital natural resource such as water can benefit from the bottom-up approach of a greater involvement of civil society and private businesses in furthering international cooperation.

References

Aggestam K (2015) Desecuritisation of water and the technocratic turn in peacebuilding. Int Environ Agreem Polit, Law Econ 15(3):327–340. https://doi.org/10.1007/s10784-015-9281-x

Allan JA, Malkawi AIH, Tsur Y (2014) Red Sea-Dead Sea water conveyance study program: study of alternatives. World Bank. https://economics.agri.huji.ac.il/sites/default/files/agri_economics/files/soa-final_march_2014.pdf. Accessed 13 Feb 2023

Anadolu Agency (2022) İsrail basını, Çavuşoğlu'nun ülkeye gerçekleştirdiği ziyareti 'tarihi' olarak niteledi. https://www.aa.com.tr/tr/dunya/israil-basini-cavusoglunun-ulkeye-gerceklestirdigi-ziyareti-tarihi-olarak-niteledi/2598189. Accessed 19 Jan 2023

Badran A (2017) Climate change and water science policy in management. In: Badran A, Murad S, Baydoun E, Dagher N (eds) Water, energy & food sustainability in the Middle East. Springer, Cham, pp 3–19

Buzan B, Wæver O, De Wilde J (1998) Security: a new framework for analysis. Lynne Rienner Publishers Inc, London, Boulder.

Cahan AJ (ed) (2017) Water security in the Middle East: essays in scientific and social cooperation. Anthem Press, London

Cook C, Bakker K (2012) Water security: debating an emerging paradigm. Glob Environ Chang 22(1):94–102. https://doi.org/10.1016/j.gloenvcha.2011.10.011

Dağ B, Rezzakoğlu A (2022) Turkiye, Israel on same page regarding normalization of ties: Turkish Foreign Minister. Anadolu Agency. https://www.aa.com.tr/en/politics/turkiye-israel-on-same-page-regarding-normalization-of-ties-turkish-foreign-minister/2597210. Accessed 19 Jan 2023.

Daily Sabah (2022) Turkey, Israel vow greater potential in endeavor to boost trade. https://www.dailysabah.com/business/economy/turkey-israel-vow-greater-potential-in-endeavor-to-boost-trade

Diriöz AO (2020) Watering down tensions: the role of securitization in water cooperation. In: Kronich S, Maghen L (eds) Ensuring water security in the Middle East: policy implications. European Institute of the Mediterranean, pp 56–79. https://www.euromesco.net/wp-content/uploads/2020/04/JPS-15-Water-Security-in-the-Middle-East.pdf. Accessed 19 Jan 2023

Diriöz AO (2021) Addressing climate change through the water–energy–food nexus: lessons learned from Turkey. In: Climate change law and policy in the Middle East and North Africa Region. Routledge, pp 136–152

Diriöz AO (2022) Promoting cooperation in natural gas development: lessons from the Balkans and the Eastern Mediterranean. In: Olawuyi DS, Pereira EG (eds) The Palgrave Handbook of natural gas and global energy transitions. Palgrave Macmillan, Cham, pp 543–563

Djernaes M (2020) Good water neighbors: rehabilitating the Jordan River through transboundary cooperation. Panorama: Solutions for Healthy Planet. https://panorama.solutions/en/solution/good-water-neighbors-rehabilitating-jordan-river-through-transboundary-cooperation. Accessed 19 Jan 2023

Dursun AE (2022) Risk analysis of natural sinkholes hazards in Karapınar basin (Konya, Turkey). Arab J Geosci 15(3):1–20. https://doi.org/10.1007/s12517-022-09564-8

Eşbah H (2021a) Peyzaj Mimarlığı Mesleği Bakış Açısından Kentsel Yağmur Suyu Hasadı. Conference Proceedings, Water Problems Workshop, June, Chamber of Mechanical Engineers, pp 85–89
Eşbah H (2021b) Suya Duyarlı Şehirler. Türkiye Water Institute. https://hetpeyzaj.com/wp-content/uploads/2022/01/Suya-Duyarli-Sehirler.pdf. Accessed 19 Jan 2023
FAO AQUASTAT (2020) http://www.fao.org/nr/water/aquastat/data/query/results.html. Accessed 19 Jan 2023
Finish Water Index (2021) Water index—an index that shows the value of water. Campaigns of the World. https://campaignsoftheworld.com/tv/finish-water-index/#:~:text=Finish%20Water%20Index%3A%20Water's%20true,poor%20country%20in%20near%20future. Accessed 19 Jan 2023
Frölich C (2020) Desecuritisation of water as a key for water diplomacy. In: Kronich S, Maghen L (eds) Ensuring water security in the Middle East: policy implications. European Institute of the Mediterranean, pp 12–13. https://www.euromesco.net/wp-content/uploads/2020/04/JPS-15-Water-Security-in-the-Middle-East.pdf. Accessed 19 Jan 2023.
Glausiusz J (2013) Environmental concerns reach fever pitch over plan to link Red Sea to Dead Sea. J Water Resour Prot-Nat 5:49–58. https://doi.org/10.1038/nature.2013.12515
Gruen GE (2007) Turkish water exports: a model for regional cooperation in the development of water resources. In: Shuval H, Dweik H (eds) Water resources in the Middle East., vol 2: Israel-Palestinian water issues—from conflict to cooperation, 157–164. Springer, Berlin, Heidelberg. https://doi.org/10.1007/978-3-540-69509-7_15
Himat A, Onüçyıldız M, Doğan S (2019) Urban drainage design according to Turkish rainwater harvesting and disposal guideline. Int J Environ Pollut Environ Model 2(4):219–226
Israel Ministry of Foreign Affairs (2022) President Herzog meets with Turkish President Recep Tayyip Erdoğan in Ankara. https://www.gov.il/en/departments/news/president-herzog-meets-with-turkish-president-erdogan-9-mar-2022. Accessed 19 Jan 2023
Kadıoğlu M (2019) Bildiğiniz Havaların Sonu: Küresel İklim Değişikliği ve Türkiye. Sia Publishing House, Istanbul
Keohane R, Nye J (2001) Power and interdependence. Longman, New York
Kibaroğlu A, Çakmak B, Doğan A (2007) Global water policies and river basin management: reflections on water resources management in Türkiye. Proceedings of the International Congress on River Basin Management, vol. II. Antalya, pp 48–58
Kirişci K, Arbell D (March 7, 2022) President Herzog's visit to Ankara: a first step in normalizing Turkey-Israel relations? Commentary, Brookings Institution. https://www.brookings.edu/articles/president-herzogs-visit-to-ankara-a-first-step-in-normalizing-turkey-israel-relations
Maghen L, Kronich S (2020) Ensuring water security in the Middle East: policy implications. Euromesco Joint Policy Study No.15. https://www.iemed.org/publication/ensuring-water-security-in-the-middle-east-policy-implications. Accessed 19 Jan 2023.
Ministry of Foreign Affairs (2004) Dışişleri Bakanlığı Sözcüsü Namık Tan'ın haftalık olağan basın toplantısı. https://www.mfa.gov.tr/4_mart-2004---disisleri-bakanlici-sozcusu-namik-tan_in_haftalik-olacan-basin-toplantisi.tr.mfa. Accessed 19 January 2023.
Mumssen YU, Triche T (eds) (2017) Status of water sector regulation in the Middle East and North Africa. World Bank Group. https://openknowledge.worldbank.org/handle/10986/27465. Accessed 19 Jan 2023.
Palestinian Investment Promotion and Industrial Estates Agency. http://www.pipa.ps/page.php?id=28878fy2656143Y28878f Accessed 13 Feb 2023.
Pradhan A, Srinivasan V (2022) Do dams improve water security in India? A review of post facto assessments. Water Secur 15:100112. https://doi.org/10.1016/j.wasec.2022.100112
Sikri R (2010) Resource competition, nationalism and their security implications. South Asian Survey 17(1):7–17. https://doi.org/10.1177/097152311001700102
Sümer V (2014) A chance for a pax aquarum in the Middle-East? Transcending the six obstacles for transboundary water cooperation. J Peacebuilding Dev 9(2):83–89

SYGM (2016) İklim Değişikliğinin Su Kaynaklarına Etkisi Projesi [Conclusion Report]. Turkish Ministry of Forestry and Water Affairs Directorate General for Water Management. https://www.tarimorman.gov.tr/SYGM/Belgeler/iklim%20de%C4%9Fi%C5%9Fikli%C4%9Finin%20su%20kaynaklar%C4%B1na%20etkisi/Iklim_NihaiRapor.pdf. Accessed 19 Jan 2023

Times of Israel—TOI. (2021). 17 June 2021. After years of delays, Jordan said to nix Red Sea-Dead Sea canal with Israel, PA. https://www.timesofisrael.com/after-years-of-delays-jordan-said-to-nix-red-sea-dead-sea-canal-with-israel-pa

Turkish Statistical Institute (TUIK—TURKSTAT) (2022). Foreign Trade Statistics, September 2022, Table 6 Top 20 Trade Countries. https://data.tuik.gov.tr/Bulten/Index?p=Dis-Ticaret-Istatistikleri-Eylul-2022-45544

UN Water (2013) What is water security? Infographic.https://www.unwater.org/publications/water-security-infographic. Accessed 19 Jan 2023

von Lossow T, Shattat M (2020) Less and less: water in the Middle East. In: Kronich S, Maghen L (eds) Ensuring water security in the Middle East: policy implications. European Institute of the Mediterranean, pp 34–55. https://www.euromesco.net/wp-content/uploads/2020/04/JPS-15-Water-Security-in-the-Middle-East.pdf. Accessed 19 Jan 2023

Willner SE, Lipchin C, Kronich S, et al (2013) A prefeasibility study on water conveyance routes to the Dead Sea. Arava Institute for Environmental Studies. https://www.researchgate.net/publication/344227332_A_PRE-FEASIBILITY_STUDY_ON_WATER_CONVEYANCE_ROUTES_TO_THE_DEAD_SEA. Accessed 19 Jan 2023

Zittis G, Almazroui M, Alpert P, et al (2022) Climate change and weather extremes in the Eastern Mediterranean and Middle East. Rev Geophys 60(3):1–48. https://doi.org/10.1029/2021RG000762

Part III
Transboundary Water Diplomacy

Chapter 8
The Evolution of Water Diplomacy Frameworks: The Euphrates-Tigris Basin as a Case Study

Ayşegül Kibaroğlu

Abstract Water diplomacy encompasses the processes and institutions through which the national interests and identities of sovereign states are represented to one another. It is enshrined in international law, which states use to explain and justify their policies to concerned actors in the international system. States mostly prefer traditional tools of water diplomacy such as negotiation and mediation to resolve disputes in transboundary river basins. This chapter explores water diplomacy along with its main principles and actors. On the one hand, the state has been the main actor in shaping transboundary water policies and conducting water diplomacy throughout the last few decades of water disputes. On the other hand, international organizations, international financial agencies, non-governmental organizations, and science-policy (Track II) initiatives also participate in water diplomacy. A brief discussion of emerging water diplomacy approaches is followed by a case study on the evolution of water diplomacy frameworks in the Euphrates-Tigris river basin.

Keywords Water diplomacy · Principles · Actors · Transboundary rivers · Euphrates-Tigris basin

8.1 Introduction

The pressure on freshwater resources, which are currently limited in many geographies around the world, is increasing with rapid population and economic growth, as well as the increase in temperatures and decrease in precipitation caused by climate change. Freshwater resources are unevenly distributed in the world, and many flow in the territories of more than one country, thus becoming transboundary water

A. Kibaroğlu (✉)
Department of Political Science and International Relations, Faculty of Economics, Administrative and Social Sciences, MEF
University, İstanbul, Turkey
e-mail: aysegul.kibaroglu@mef.edu.tr

K. Szálkai and M. Szalai (eds.), *Theorizing Transboundary Waters in International Relations*, Springer Water, https://doi.org/10.1007/978-3-031-43376-4_8

resources. Transboundary river basins are home to 52% of the world's population, cover 47% of the land area, and they account for approximately 60% of global water flow (McCracken and Wolf 2019: 732–782). As countries with transboundary water resources allocated these resources for socio-economic development, increasing the domestic water, food, and energy needs of the population in both urban and rural areas since the Industrial Revolution, disputes have emerged between countries located in the same basin regarding the use, management, and allocation of transboundary water resources.

The comprehensive security approach that developed with the end of the Cold War brought the phenomenon of water disputes to the fore in the context of environmental security (Myers 1993). In this context, the question was raised whether water scarcity will be the cause of interstate conflict (i.e. water wars), and opinions have been put forward about the increasing competition over transboundary water resources, which could turn into hot conflicts. The first part of this chapter explains the main propositions of the water wars literature, arguing that the main premises of the water wars approach are insufficient to explain interstate transboundary water relations in the current situation, and empirical evidence shows that states prefer legal and political tools of diplomacy instead of engaging with conflict to handle water disputes.

In dealing with transboundary water disputes, states mostly prefer water diplomacy mechanisms such as negotiation, mediation, conciliation, and fact-finding commissions. For decades, when water-related disputes arose, the state was the main actor in determining transboundary water policies and conducting water diplomacy. State bureaucracy encompasses diplomats, the legal advisers of foreign ministers, and technocrats responsible for the development and management of the water resources in the relevant ministries of water, irrigation, environment, forestry, and energy. The discourses and practices of the states were shaped and sometimes transformed during the negotiations involving the long-term water disputes, which has played an important role in changing the structure of transboundary water relations (Kibaroglu and Scheumann 2013: 279–307). States sometimes received loans from international credit agencies (e.g. World Bank) and development agencies (e.g. U.S. Agency for International Development—USAID) to construct high-cost water infrastructure (dams, irrigation canals) on transboundary waters. These funding agencies thus became involved actors in the water allocation negotiations through the conditions they attached to the loan agreements (Gürün 1994: 240–272). On the other hand, international organizations, international non-governmental organizations, and science-policy initiatives (Track II diplomacy) joined water diplomacy as new actors as the issue of scarcity, pollution, and the sharing of water resources came to the attention of the international community. With the participation of these actors in water diplomacy, new approaches related to the management of transboundary water resources, for example, the sharing of benefits such as energy, food, and services to be obtained from water resources instead of sharing water resources per se, have been developed. These emerging new approaches and actors of water diplomacy are also discussed in this chapter.

With the increase in the use of transboundary rivers for consumption purposes, for example, the expansion of irrigation since the last quarter of the nineteenth century, many states have claimed mutual sovereignty and right of use over transboundary water resources. As a result of these discourses and actions, a series of international customary law principles have been adopted. The riparian states of transboundary waters have signed numerous bilateral and multilateral agreements, protocols, and memoranda of understanding on the use, management, and sharing of these resources. In other words, treaty (written) and customary international water principles of law, as well as traditional diplomatic methods have been developed by the states as the main tools for resolving disputes regarding transboundary waters. Hence the chapter describes the role of international water law principles in shaping water diplomacy practices.

Certain regions in the world are among the most problematic ones in terms of transboundary surface and groundwater resources management and allocation between two or more countries. The Middle East is regarded as one of the most challenged regions in this regard. In addition to the constraints of natural water resources, the region suffers from an abundance of issues that compound water security, including a rapidly growing population, uneven economic development, a limited amount of water supply that is irregularly distributed, the negative impacts of climate change and variability, and poor water management and allocation practices both within and between states. Some 60% of the water in the region flows across international borders, complicating resource management. The geopolitical importance of the region and the conflicts that have consequently resulted aggravate the usual problems of using water in a variety of settings, e.g. in the Euphrates-Tigris (ET) basin. The last part of the chapter presents the evolution of water diplomacy institutions and processes in the ET basin in the form of a case study.

8.2 The Rise of the Water Issue in International Politics and the Water Wars Approach

With the end of the Cold War, the concept of security started to be discussed at the individual, state, and international system level, which all interact with each other, approaching it from a comprehensive perspective, where not only political and military threats but also economic, social, and environmental factors are considered (Myers 1989: 23–41; Mathews 1989: 162–177). In the post-Cold War context, the issue of water ceased to be a purely low-politics, technical issue and moved to the realm of high-politics security issues, especially in the Middle East, Africa, and South Asia, which are located in semi-arid and arid climatic zones, where water resources are very limited. In fact, by the end of the 1980s, foreign policy-making institutions in riparian countries in transboundary settings started to discuss the issue of transboundary waters both in their bilateral political relations with their neighbors and around processes of multilateral political dialogue.

During the 1980s and 1990s, the possibility of international hot conflicts regarding the sharing of water resources was asserted by the water wars literature as a logical consequence of the scarcity of water resources. The water wars approach, especially with regard to the water issues in the Middle East, claims that water is becoming an increasingly scarce resource in the region, and it will be the primary cause of a dispute that may result in armed conflict (Cooley 1984: 3–26). The pioneers of this approach preferred slogan-style titles such as "coming water wars" in their works, and they referred to the harsh statements made by the political leaders of riparian countries regarding transboundary water issues as proof of the escalation of conflict among pivotal riparian states, for example, Egypt and Ethiopia, or Israel and Jordan (Bullock and Darwish 1993). However, the political rhetoric used by leaders at the domestic level have been usually in contradiction with their foreign political and economic policies towards their neighbors.[1]

This approach also draws attention to the fact that in regions such as the Middle East, where there are many complex political disputes and conflicts, it becomes difficult to separate water-related problems from the deeply rooted political problems if there is a disagreement between riparian countries regarding water sharing. M. Lowi (1993) analyzes the dispute in the Jordan river basin in this context, where Arab–Israeli relations have both water problems and broader political conflicts. Lowi argues that in this environment of deep historical conflicts, the problem faced by countries in transboundary settings is not only a water problem, it is also related to many other aspects of the intricate problems between the states (Lowi 1993). In other words, water disputes can become more complex between countries that already have problematic relations that require resolution, and they can reach the level of high politics (Bakour and Kolars 1994). On the other hand, during the diplomatic negotiations that took place within the scope of the Middle East Peace Process, the parties used the creative feature of diplomacy by addressing the conflict regarding water resources in the Jordan basin both among the main problems (e.g. borders, sovereignty, refugees, the status of Jerusalem) and the functional issues that support the solution of the main problems (e.g. disarmament, economic cooperation, environmental protection), in addition to transforming water into a field that supports the negotiation processes of both strategic and functional issues (Libiszewski 1997: 385–402).

Even if the expectations of the water wars approach centers around conflictual events, this approach can still explain and understand cooperation, but only in a limited way, under certain circumstances. Thus, the water wars approach refers to certain conditions for cooperation between countries: the existence of a hegemonic power, which can lead to basin-wide cooperation, as well as the hegemon undertaking to create such cooperation are the most important of these conditions. In other words, cooperation in transboundary water basins is possible only if the hegemonic power in the basin is willing to cooperate (Waterbury 1991). By making sharp distinctions between the political, military, economic, and geographical acquisitions of the

[1] For instance, in the early 1990s, King Hussein of Jordan emphasized the fact that the Kingdom of Jordan could wage another war with Israel only because of the issue of water sharing, although he was engaged in secret negotiations (so-called 'picnic table talks') that later on culminated in the Peace Treaty signed with Israel in 1994, which includes water sharing clauses. See Kibaroglu (2002).

riparian states, the water wars proponents support that upstream countries are in the most advantageous position. They argue that even if negotiations begin, the downstream countries will be at a disadvantaged position at the bargaining table. Based on this argumentation, the water wars literature does not see any prospects for a fruitful outcome through a negotiation process, largely because the least needy and/or most powerful riparian state will derive little benefit from cooperating and relinquishing its most favorable position. Nevertheless, the water wars approach also argues that certain states that could act as a hegemon, relying on their military and economic power, might be neither sufficiently powerful to take on the role of a hegemon, nor would they have any incentive to do so (Waterbury 1991). From this point of view, this approach denies the possibility of a solution based on fair and balanced negotiation processes.

A research team led by A. T. Wolf at Oregon State University studied sequences of international conflicts that include existing armed conflicts and ones at risk of emerging due to the sharing of water resources (Brecher and Wilkenfeld 1997). Their systematic study shows that the historical reality is quite different from what the water wars literature claims. In modern history, only seven mild hot conflicts have occurred regarding transboundary water resources, and these conflicts have many main causes other than the desire to dominate water resources (Wolf 2013). In fact, states seem to prefer to carry out intensive and long-term negotiation processes for the resolution of disputes instead of engaging in hot conflict, and many agreements have been concluded as a result of these negotiations (FAO 1984). Diplomatic methods have played an undeniable role in the way these negotiations were conducted and the agreements were realized.

8.3 Water Diplomacy: Basic Characteristics and Actors

Water diplomacy is developing as a relatively new field compared to the traditional fields of diplomacy, which focus on political, military, and economic relations. In its most basic definition, water diplomacy covers the contacts, consultation mechanisms, and systematic negotiations between the official representatives of the states (i.e. politicians and diplomats) to reach legally binding documents (e.g. treaties, agreements, protocols, or memoranda of understanding) or non-binding soft law documents (e.g. action plans, guidance documents, or proposals) in order to solve water problems that cross borders.

Water diplomacy is an area where the principle of reciprocity as the basic principle of traditional diplomacy is evidently valid between riparian states. A state can be an upstream riparian in one transboundary river basin and a downstream one in another transboundary river basin. Thus, riparian states, regardless of their geographic position, be it upstream or downstream, are inclined to adopt agreeable positions in their cross-border water relations so that they can expect similar positive policy responses from their counterparts. In accordance with this basic principle of reciprocity, whenever riparian states have disputes over water resources, they opt for utilizing the basic methods of water diplomacy negotiations.

International institutions are entities that regulate interstate relations and support the development of cooperation processes (Keohane 1993: 271). Through international institutions, actors that have common interests in the system tend to cooperate. The existence of these institutions increases both the scope and the rate of information exchange on issues of dispute among competing countries, and they support the development of harmonious interests among states with the coordination mechanisms established by these countries. One of the main goals of water diplomacy is the establishment of institutions such as joint technical committees, joint water committees, or river basin organizations. With the facilitation and support of water diplomacy mechanisms, institutions, if properly constructed, can provide predictability in transboundary water use, coordinate interstate water development policies, and pave the way for improved management and the allocation of transboundary water resources. These institutions may ensure the continuity of negotiations in transboundary waters, and they can be used as a mutual regular consultation and negotiation platform to reach legally binding bilateral or multilateral agreements, in addition to being responsible for the implementation and monitoring of water agreements. If the issue of transboundary water management is considered as a part of economic welfare issues among riparian countries, the potential for cooperation is strengthened. Thus, water diplomacy mechanisms become instrumental in building cooperation between riparian states in technical fields such as the exchange of hydrological information, flood and drought management, and the establishment of joint hydroelectric and water treatment facilities. Moreover, growing technical cooperation may also attract the attention of international funding institutions, which might offer technical and financial support to riparian states.

Since the first quarter of the twentieth century, when transboundary water problems began to be experienced intensely, the state has become the main actor in water diplomacy. In bilateral or multilateral transboundary water negotiations, the relevant diplomats from the foreign ministries, legal advisers, and senior representatives and technocrats from the ministries responsible for water resources development and management have been the main actors directly conducting water negotiations between countries (Kibaroglu and Ünver 2000: 311–330).

On the other hand, when developing countries seek technical and financial support from the international funding agencies for large dams and irrigation projects on transboundary rivers, these actors also become influential actors on transboundary water policy. For instance, Turkey agreed to take all necessary measures to ensure the release of 350 m^3/second of water from Keban Dam to downstream Syria as a result of the agreement it signed with the United States Agency for International Development-USAID during the construction period of the dam on the Euphrates River. This situation was confirmed to the Syrian and Iraqi sides in the same year (Gürün 1994: 240–272). The 350 m^3/second application implemented with this agreement was later on turned into the delivery of 450 m^3/second water from the dam downstream as a precondition for a loan from the World Bank during the construction of Karakaya, the second largest dam Turkey built on the Euphrates in the mid-1970s. In this case, international funding agencies and other relevant third parties attached the credit, grant, and support policies to the conditions, and they imposed obligations primarily on the upstream country (Turkey) and only protected the rights of the downstream country. This has caused some upstream countries like Turkey to take a cautious and distant stance towards third-party involvement in water diplomacy.

By the post-World War II period, international funding organizations and development agencies aimed to reach a compromise between parties in disputed transboundary river basins within the framework of their more balanced and cooperation-oriented policies, in order to support negotiations and create a problem-solving atmosphere. They also tried to make transboundary water agreements more attractive by providing financial support. To illustrate, the World Bank played a key role in the realization of the water sharing agreement between India and Pakistan regarding the Indus River (Alam 2002). Similarly, the UN Development Program and the World Bank played a significant role in the Mekong river negotiations, which led to the establishment of the Mekong River Committee and later the Mekong River Commission, which is mainly composed of the downstream riparian countries.

Diplomacy is no longer only conducted by politicians and diplomats. Along with the practices of official diplomats, civil society, professionals, and academia play a significant role in building trust and cooperation among otherwise competing states (Kibaroglu and Sayan 2021). Track I diplomacy is typically carried out by government officials, who use bargaining, negotiation, and other peaceful means to negotiate treaties and accomplish international agreements. On the other hand, expert groups consisting of academics, diplomats, bureaucrats, technocrats, and professionals (called Track II informal diplomacy) are accepted as authority in terms of knowledge and experience in their fields of expertise, and they sometimes take the initiative to pave the way for interstate relations and support the negotiation processes in certain areas with new and constructive ideas and projects based on scientific data and experience, ultimately supporting various cooperation frameworks. Track III diplomacy, in turn, refers to the use of non-traditional actors, including non-governmental organizations, business executives, and private citizens, who typically conduct dialogue and problem-solving activities in their own domains of interest (Kibaroglu and Sayan 2021).

The Euphrates Tigris Initiative for Cooperation (ETIC), an example of Track II initiatives, started its activities in 2005 as a cooperation initiative of experts from the riparian countries of the Euphrates-Tigris basin. Scientists and experts working in various fields of water resources management came together within the framework of ETIC and formed an interdisciplinary group. Within a framework of workshops addressing the current situation and the water policies in the Euphrates-Tigris basin, including the needs, priorities, and expectations of the riparian countries, the group made progress within a short period of time in defining common problems and producing common approaches to solve these problems, acting like the epistemic communities defined by Peter Haas (1992).

ETIC defined its mission and vision at its founding meeting in 2005, where the founding members emphasized that they would act together for development and cooperation within the political environment offered by the current conditions, needs, and opportunities in the Euphrates-Tigris basin. ETIC follows a holistic approach with its development-oriented, multi-sectoral activities. Its members, who share the view that discussions focusing only on water sharing are inconclusive and lead to divergences rather than integration, aim to set a new, comprehensive cooperation agenda, which can result in benefits for all stakeholders in the basin. However, ETIC does not seek to impose a water-sharing formula or any model of cooperation, instead it tries to create an environment that can enable dialogue between the parties. ETIC continues its various activities with the principles of impartiality and facilitation within the framework of the program areas it defines as water and land resources, water resources management and socio-economic development (Kibaroglu 2008: 191–195).

ETIC aims to create dialogue for the purpose of realizing socio-economic regional development among riparian countries in the basin, which is not far from the official intergovernmental dialogues; in fact, ETIC carries out activities that encourage and support these initiatives. By taking advantage of international water congresses such as the World Water Forum, the World Water Congress or World Water Week, ETIC has the opportunity to share the mission, vision, and goals of the initiative with the international community, which consists of academics, professionals, technocrats, and politicians. By succeeding in bringing together the officials responsible for water resources policy and the management of riparian states on the same platforms, it supports cooperative dialogue for sustainable development in the Euphrates-Tigris basin.

The growing variety of actors in water diplomacy has provided opportunities for riparian states to start or revitalize dialogues, contacts, and negotiations on non-contentious, promising fields of cooperation, e.g. water-based socio-economic development or better water governance through inclusive and efficient water management institutions, particularly through the involvement of professionals, academia, business, and civil society in water diplomacy practices.

8.3.1 The Role of International Water Law in Water Diplomacy

International water law plays an important role in resolving water-related disputes and determining the universal rules for better management and allocation of international water resources. Thus, international water law experts work on three main sources of international law: bilateral and multilateral treaties on international water resources, customary international law rules developed in the framework of the codification of water law, by the activities of international organizations, the UN International Law Commission (ILC), and the International Law Association (ILA), which is an independent professional organization, and the doctrines that have developed through a process of claims and counter-claims between riparian states on a transboundary river

(Garretson et al 1967; Caponera 1985; McCaffrey 1991a). Based on their experience regarding the legal aspect of water issues, international water law experts emphasize the fact that the unique characteristics of international water law can provide the main principles and rules that will pave the way for states to establish effective water diplomacy mechanisms on transboundary waters.

International water law provides a series of principles and norms regarding the settlement of transboundary water disputes; however, in many cases it has not been possible to transform these principles into effective institutions for the management of transboundary water relations (Dellapenna 1995). Water diplomacy practitioners, namely decision-makers and diplomats should take an active role in implementing these teachings at the level of principles and norms developed by the international law experts.

Stephen C. McCaffrey (1993) underlines the fact that international law is a non-supranational system, that is to say, its implementation is only possible with the consent of the states, which in fact create the basic principles and norms of international law. Compared to the national law system, international law lacks some basic features, such as compulsory jurisdiction and the related enforcement mechanisms. Despite this weak structure of the international legal system, states mostly try to prove that they have acted in accordance with the norms of international law and accepted the international legal system as a basic frame of reference (McCaffrey 1993: 98).

Many principles of international water law are fed by two types of sources: treaties and customary international law. Rules resulting from treaties are relatively easy to identify, although there is always the possibility that some articles may be interpreted differently. In a systematic historical analysis, the UN Food and Agriculture Organization (FAO) identified 3,600 agreements on water resources between 805 and 1984 (FAO 1984). In addition, the full text of 149 treaties on the use of transboundary waters only, excluding those that concern boundary waters and fishing rights, has been compiled by experts from Oregon State University in the Transboundary Freshwater Dispute Database. This compilation has created a valuable source that can be used by researchers interested in studying transboundary water disputes, areas of cooperation, or more generally, international water law.

However, it is more difficult to establish and implement customary international law rules, although the codification efforts of the leading international water law institutions regarding these rules have contributed greatly to this process. Even though the role of law in resolving disputes changes, states rarely violate the accepted principles of international law, in fact, they often rely on these rules in their diplomatic relations. According to S. C. McCaffrey (1991b), the role of these norms and rules in the resolution of international water disputes will increase as applicable legal principles become more concrete and more generally accepted. In this context, international law experts emphasize the fact that the adoption of the United Nations Convention on the Law of Non-Navigational Uses of International Watercourses in the UN General Assembly in 1997 was a remarkable achievement in the codification and prospective development of (customary) international law rules. The Convention entered into force on August 17, 2014, after obtaining the sufficient number of ratifications (35 countries).

Some of the main principles of the 1997 UN Convention include the right to equitable use and the obligation not to cause significant harm, the duty to cooperate, and regular exchange of hydrological and other relevant data and information. Undoubtedly, these principles provide useful references for riparian states that wish to reach agreements in transboundary river basins. The leading principle of international water law, the principle of equitable and reasonable utilization and participation, and more generally its fair and reasonable use, defines the development and protection of international watercourses, thereby drawing up the framework for the actions that need to be taken. Riparian states should make these general principles applicable, measurable, and verifiable through the rules and decision-making methods provided by international treaties related to the specific conditions of transboundary water resources.

However, apart from the legal and legitimate grounds of international water law and water diplomacy, such as conducting negotiations and establishing and applying international legal principles, 'coercive diplomacy' tools are also sometimes used on water, especially in the transboundary river basins in the Middle East. Coercive diplomacy includes techniques such as the threat of using force in foreign policy, ultimatum, implied ultimatum, gradual pressure, or try and see. One typical example is from the early 2000s, when Israel declared any change by Lebanon in the Wazzani River *casus belli* (Zeitoun 2007). The basic logic is that the coercive actor forces its opponent to make a choice by expressing or threatening to use force (George 1991).

8.3.2 Innovative Approaches in Water Diplomacy

In addition to the legally binding customary and treaty law norms, water diplomacy actors have adopted and implemented new approaches that can be evaluated within non-binding soft law. In this context, the benefit sharing concept is introduced by the World Bank experts (Sadoff and Grey 2002: 389–403) with an aim to facilitate sustainable development and water cooperation in transboundary river basins such as the Nile, Orange-Senqu, Limpopo, Okavango, Zambezi, Incomati, and Congo (Turton 2008: 180–200). In line with this approach, various benefits can be identified in a transboundary river basin, thus, the purpose is to share the benefits obtained from water, not water itself. Indeed, in transboundary river basins, establishing close links between water resources management and development-related sectors and resources will allow for the creation of a regional benefit pool. Thus, the agenda of cooperation targeted in transboundary river systems will expand, and the necessary link will be established between cooperation and regional development. In transboundary river basins, states may prefer to adopt benefit sharing approach, which incorporates coordinated policy and practices in the water-based development sectors such as energy, agriculture, health, and environment. The benefit-sharing approach could also facilitate stability, peace, cooperation, and prosperity in transboundary river basins (Table 8.1).

Many studies have been carried out in different disciplines to find solutions to the problems that arise between states regarding transboundary waters. One of these efforts is the Water Diplomacy Framework approach (Islam and Susskind 2013: 1–41). This approach emphasizes the fact that water problems become more complex as a result of crossing legal, physical, and disciplinary boundaries. It suggests that traditional engineering techniques and economic approaches cannot solve water problems and advocates the development of joint research and the problem-solving abilities of actors that have a say in water management through open-ended negotiations. Since 2011, the pioneers of this approach have been organizing water diplomacy workshops every year, training the participants in diplomatic negotiation, aiming to bring together future water negotiators and contribute to the creation of a global network where the participants can share their experiences. The water diplomacy framework approach argues that new values can be created by parties coming to the negotiation table equipped with the power to evaluate different options instead of fixed and unchanging negotiation techniques (Islam and Madani 2017).

Table 8.1 Potential benefits of regional water cooperation

Potential benefits of regional water cooperation		
	Economic activities	Other activities
Benefits resulting from good water governance	**Economic benefits** Broad activity and productivity in economic sectors (irrigated agriculture, hydropower, industry) Reducing the cost of carrying out productive activities Reducing the economic damage of water-related disasters (floods, droughts)	**Social and environmental benefits** Positive contribution of good water quality and reducing the risk of water-related disasters for the health sector Positive impact of economic benefits for employment and poverty reduction Improved access to services such as electricity and water Preventing environmental degradation and loss of biodiversity
Benefits resulting from an environment of comprehensive trust	**Benefits of regional economic integration** Development of regional markets for products, services, and the workforce Developing and increasing foreign investments and infrastructure networks	**Benefits for peace and security** Avoiding the cost of military conflicts Savings from reducing military spending Other geopolitical benefits

8.4 Water Diplomacy Mechanisms in Practice: The Euphrates-Tigris Basin as a Case Study

Transboundary water issues started to be a part of regional politics when the three major riparian states of Turkey, Syria, and Iraq[2] introduced major water and land resource development projects in the Euphrates-Tigris (ET) basin. In this context, large-scale dams and irrigation systems were initiated by the early 1960s. Owing to the competitive nature of these uncoordinated national water development projects, disagreements over transboundary water uses surfaced in the late 1960s, and the riparian states opted for diplomatic negotiations to deal with their disagreements. The main topic of these negotiations was the impact of the construction of the Keban Dam in Turkey as well as that of the Tabqa Dam in Syria on Iraq's historical water use patterns. As the national water development projects progressed, incompatibilities between water supply and demand occurred throughout the river basin. Thus, the water wars approach indicated that the ET basin was one of the hot spots for the "coming water wars" due to the rising competition over the use of water resources by the three major riparian states. Even though no hot conflict was reported among the riparian states concerning water sharing, the sporadic technical negotiations could not prepare the ground for a comprehensive treaty on equitable and effective transboundary water management in the basin. Hence, a series of diplomatic crises occurred in the region during the last quarter of the twentieth century (Kibaroglu and Scheumann 2013).

In the ET basin, the riparian states preferred water diplomacy mechanisms, namely diplomatic negotiations to resolve the crises (Kibaroglu and Ünver 2000). Most of the crises were related to Iraq's concerns regarding the impact of the construction and the filling of the dams in Turkey and Syria. Thus, diplomats and technocrats from the three countries met several times, although on an irregular basis, to exchange information concerning the technical details of the construction and the filling of the dams. In the decades since the disputes over water first began, the state has been the major actor in the formulation and implementation of water diplomacy in the ET basin. Water negotiations were held by technocrats from the central water agencies of the riparian states, accompanied by diplomats who advised and monitored the negotiations, particularly when international legal and political aspects were under discussion (Kibaroglu and Scheumann 2013). The discourse and practices of the state bureaucracies, water technocrats from various ministries, and foreign office diplomats evolved during the prolonged water dispute. In the period between the 1960s and the 1990s, the riparian states were too rigid in their position, emphasizing their absolute water rights over the rivers. With the emergence of a conducive

[2] The Euphrates and its tributaries drain an enormous basin of 444,000 km^2, of which 33% lies in Turkey, 19% in Syria, and 46% in Iraq, while the Tigris and its tributaries drain an area of 387,600 km^2, of which 15% lies in Turkey, 0.3% in Syria, 75% in Iraq, and 9.5% in Iran. Both rivers rise in Turkey, scarcely 30 km apart, flow through Syria and Iraq, and join to form the Shatt-al-Arab waterway north of Basra in Iraq, before discharging into the Persian Gulf. See Kibaroglu and Scheumann (2013: 282).

overall political environment in the early 2000s, state representatives adopted a more needs-based approach by concluding a series of memoranda of understanding on the protection of the environment, water quality management, water efficiency, drought management, and flood protection, with a view to addressing the adverse effects of climate change (Kibaroglu and Scheumann 2013).

Even though the riparian states have opted for bilateral and trilateral diplomatic negotiations for settling disputes, there still does not exist a basin-wide agreement for the effective and equitable use and management of transboundary waters in the ET basin. In 1987 and 1990 two bilateral protocols, acknowledged by all the riparian states as being interim agreements, were signed following a number of high-level meetings by top officials. Turkey and Syria signed the Protocol on Economic Cooperation in 1987, which contained provisions related to the allocation of the waters of the Euphrates River. According to the Protocol, Turkey guaranteed to release 500 m^3/second from the Euphrates, with deficiencies in any month to be compensated the following month (Protocol 1987).

In 1989, Turkey had to interrupt the flow of the Euphrates for several weeks when the Atatürk Dam reservoir was being filled. This caused anxieties on the Syrian and Iraqi sides, and they agreed to determine their bilateral shares from the Euphrates before such an interruption should occur again as Turkey's Southeastern Anatolia Development Project (GAP in Turkish) progressed. Therefore, a bilateral agreement between Syria and Iraq was signed on April 16, 1990, at the 13th meeting of the Joint Technical Committee (JTC) in Baghdad, according to which 58% of the Euphrates waters coming from Turkey would be released to Iraq by Syria. The Protocol stipulates that "The contingent of water to Iraq passing through the Syrian-Iraqi border is to be a permanent annual total rate of 58 per cent of the river water passing into Syria at the Syrian-Turkish border. The Syrian contingent of the river waters is to be the rest of the waters, totalling 42 per cent of the waters passing through the Syrian-Turkish border" (Law No. 14 1990). However, both water protocols were poorly crafted, with various shortcomings and loopholes in addressing the fundamental complexities in transboundary water management, such as the handling of water quantity and quality issues concomitantly with the protection of environmental resources, as well as addressing variability in the flow and providing for adjustments under the impact of climate change.

On the other hand, in the early 1980s, the riparian states in the Euphrates-Tigris basin managed to establish the institutional framework of the Joint Technical Committee, members of which included participants from all three riparian states. However, the riparian states did not agree to give the JTC clear and commonly agreed functions. On the contrary, the states continued unilateral and uncoordinated water and land development projects, and the JTC meetings did not make an effective contribution to the settlement of the transboundary water dispute. It also did not provide a platform for delineating the priorities and needs of the co-riparians as a basis for addressing regional water problems (Kibaroglu and Scheumann 2013).

Water diplomacy mechanisms, particularly at the transboundary level, have been introduced in the ET basin with the aim to reach agreeable solutions between parties that had diverging interests as well as competing water development schemes. With the help of formal institutions like the JTC, high-level water diplomacy frameworks, water sharing protocols, and memoranda of understanding, the national interests and identities of sovereign states are represented to one another. Although these institutions may not have been effective most of the time in terms of the protection and efficient use and management of water and other related resources, they have served to place transboundary water issues within a legitimate and peaceful realm rather than mixing them with potentially conflict-laden issues, such as border security and territorial disputes, which might otherwise escalate into hot confrontations.

Some of these institutions comprise organizational structures, while others represent intangible principles and rules contained in bilateral protocols and memoranda of understanding. The formal transboundary organizational structures in the basin, of which the JTC is the most significant one, have failed to be active all the time, and they have also become ineffective in responding to the concerns and needs of the riparian states. Other institutions, for example, bilateral water allocation treaties, face various challenges particularly in relation to sustainable management and the protection of transboundary water resources under the growing impact of climate change.

8.5 Conclusion

Water diplomacy is the art of handling transboundary water disputes before they turn into hot conflicts by using the tools of diplomacy, for example negotiations and international water law. However, despite all the attempts of water diplomacy, transboundary water issues continue to be the subject of political confrontations and prolonged negotiations between parties, and final and permanent water agreements have been achieved in very few transboundary water basins. Water diplomacy has tried to create new areas of cooperation that will allow to improve cooperation between riparian states. These initiatives, which deal with the transboundary water issue on the basis of regional socioeconomic development and cooperation, are carried out in various basins around the world. The Nile Basin Initiative, the South Africa Development Community (SADC), and the Mekong River Commission are the leading ones among these initiatives. The parties that took advantage of the end of the Cold War and the end of the division in Europe have discussed water-based cooperation around advanced issues such as the protection of water resources and improving their quality, and they have arrived at a point where they can make decisions that bind state institutions with the participation of not only state representatives but also a wide range of stakeholders from different sectors. On the other hand, the ET basin case study in the Middle East demonstrates that even in the absence of harmonious relations and political stability, various institutions among concerned actors could be established in the ET basin. The riparian states have maintained contact over the ET basin at different levels by establishing and revitalizing governance mechanisms such as the JTC, memoranda of understanding, as well as informal Track II diplomacy initiatives. During these processes, multiple actors, ranging from bureaucracies and heads of state to unofficial actors such as ETIC, have focused on cooperative socio-economic development problems over divisive issues of water sharing in the basin, which has proven to be progressive and facilitative in sustaining cooperation in the basin.

References

Alam UZ (2002) Questioning the water wars rationale: a case study of the Indus Waters Treaty. Geogr J 168(4):341–353

Bakour Y, Kolars J (1994) The Arab Mashrek: hydrologic history, problems and perspectives. In: Rogers P, Lyndon P (eds) Water in the Arab world: perspectives and prognoses. Harvard University, Cambridge, pp 121–146

Brecher M, Wilkenfeld J (1997) A study of crisis. University of Michigan, Ann Arbor

Bulloch J, Darwish A (1993) Water wars: Coming conflicts in the Middle East. Víctor Gollancz, London

Caponera D (1985) Patterns of cooperation in international water law: principles and institutions. Nat Resour J 25(3):563–587

Cooley JK (1984) The war over water. Foreign Policy 54:3–26

Dellapenna JW (1995) Building international water management institutions: the role of treaties and other legal arrangements. In: Allan JA, Mallat C (eds) Water in the Middle East: Legal, political and commercial implications. I. B. Tauris, London, pp 55–89

ETIC (2020) Activities. https://euphratestigrisinitiativeforcooperation.wordpress.com/activities. Accessed 5 July 2019

FAO (1984) Systematic index of international water resources treaties, declarations, acts and cases, by basin. Vol II. https://www.fao.org/3/AK470T/ak470t.pdf. Accessed 18 July 2022

Garretson AH, Hayton RD, Olmstead CJ (1967) The law of international drainage basin. Oceana, Dobbs Ferry

George A (1991) Forceful persuasion: coercive diplomacy as an alternative to war. United States Institute of Peace, Washington, DC

Gürün K (1994) Akıntıya kürek çekmek: bir büyükelçinin anıları. Milliyet, İstanbul

Haas P (1992) Introduction: epistemic communities and international policy coordination. Int Organ 46(1):1–35

Islam S, Susskind L (2013) Water diplomacy: a negotiated approach to managing complex water networks. Routledge, New York

Islam S, Madani K (eds) (2017) Water diplomacy in action: contingent approaches to managing complex water problems. Anthem Press, London-New York

Keohane RO (1993) Institutional theory and the realist challenge after the Cold War. In: Baldwin D (ed) Neorealism and neoliberalism: the contemporary debate. Columbia University, New York, pp 269–300

Kibaroglu A (2002) Building a regime for the waters of the Euphrates-Tigris River Basin. Kluwer Law International. London-The Hague-New York

Kibaroglu A (2008) The role of epistemic communities in offering new cooperation frameworks in the Euphrates-Tigris Rivers System. J Int Aff 61(2):183–198

Kibaroglu A, Sayan RC (2021) Water and 'imperfect peace' in the Euphrates-Tigris river basin. Int Aff 97(1):139–155

Kibaroglu A, Scheumann W (2013) Evolution of transboundary politics in the Euphrates-Tigris River System: new perspectives and political challenges. Glob Gov 19(2):279–305

Kibaroglu A, Ünver O (2000) An institutional framework for facilitating cooperation in the Euphrates-Tigris River Basin. Int Negot 5(2):311–330

Law No. 14 of 1990, ratifying the Joint Minutes concerning the provisional division of the waters of the Euphrates River between the Republic of Iraq and the Arab Republic of Syria. http://www.cawater-info.net/bk/water_law/pdf/euphrates_e.pdf. Accessed 31 Jan 2023
Libiszewski S (1997) Integrating political and technical approaches: lessons from the Israeli-Jordanian water negotiations. In: Gleditsch NP (ed) Conflict and the environment. NATO ASI Series, vol 33. Springer, Dordrecht, pp 385–402
Lowi M (1993) Water and power: the politics of a scarce resource in the Jordan River Basin. Cambridge University Press, New York
Mathews JT (1989) Redefining security. Foreign Aff 68(2):162–177
McCaffrey SC (1991a) International organizations and the holistic approach to water problems. Nat Resour J 31(1):139–165
McCaffrey SC (1991b) The evolution of the law of transboundary rivers. Paper presented at the Prospects for Regional Cooperation on Transboundary Waters in the Middle East international conference, Bilkent University, Ankara, 1991
McCaffrey SC (1993) Water, politics, and international law. In: Gleick PH (ed) Water in crisis: a guide to the world's fresh water resources. Oxford University Press, New York, pp 92–99
McCracken M, Wolf AT (2019) Updating the register of international river basins of the world. Int J Water Resour Dev 35(5):732–782
Myers N (1989) Environment and security. Foreign Policy 74:23–41
Myers N (1993) Ultimate security: the environmental basis of political stability. WW Norton, New York
Protocol on matters pertaining to economic cooperation between the Republic of Turkey and the Syrian Arab Republic (1987) United Nations Treaty Series 87/12171, 17/7/1987
Sadoff CW, Grey D (2002) Beyond the river: benefits of cooperation on international rivers. Water Policy 4(5):389–403
Turton A (2008) A South African perspective on a possible benefit-sharing approach for transboundary waters in the SADC region. Water Altern 1(2):180–200
Waterbury J (1991) Dynamics of basin-wide cooperation in the utilization of the Euphrates. Paper presented at The Economic Development of Syria: problems, Progress, and Prospects Conference, Damascus, 6–7 Jan 1991
Wolf AT (ed) (2013) Transboundary freshwater dispute database. Oregon State University. http://www.transboundarywaters.orst.edu. Accessed 27 July 2020
Zeitoun M (2007) Violations, opportunities and power along the Jordan River: security studies theory applied to water conflict. In: Shuval H, Dweik H (eds) Water resources in the Middle East: Israel-Palestinian water issues—From conflict to cooperation. Springer, Berlin-Heidelberg, pp 213–224

Chapter 9
Is Salmon Surprise on the Table? Indigenous Rights and Multi-Stakeholder Diplomacy in the Renegotiation of the Columbia River Treaty

Kinga Szálkai and Mary Durfee

Abstract As a consequence of hydropower utilization and environmental change, the number of salmonids in the Columbia River Basin shared by the US and Canada declined sharply through the twentieth century. This meant a great economic and cultural loss for the Indigenous communities of the basin, especially as they could not participate in the decisions concerning their lands, societies, and future. The current renegotiation of the 1961 Columbia River Treaty, regulating flood control and hydropower production between the two states, offers a historic opportunity to address these problems. In order to modernize the Columbia River Treaty, however, a new understanding of transboundary water diplomacy is necessary. The authors argue that the renegotiation process needs to go beyond multi-stakeholder approaches and embrace the sovereignty of the Indigenous partners. The salmon issue, should it ultimately be addressed, could prove transformative for the relationship between Indigenous people and their tribal governments and other sovereigns, and it would mark a significant stride towards reconciliation and decolonizing policy in the US and Canada.

Keywords Columbia River Treaty · Indigenous people · Salmon · Multi-stakeholder diplomacy · Sovereignty · Ecosystem

---

The author Mary Durfee lives in the Columbia River Basin, about a mile from the Willamette River, and clearly benefits from the current arrangement for power and flood control. She is also in the territory of the Kalapuya, and she extends her respect and gratitude to the many Indigenous people who call these lands home.

K. Szálkai (✉)
Budapest Metropolitan University, Budapest, Hungary
e-mail: kszalkai@metropolitan.hu

M. Durfee
Michigan Technological University, Houghton, MI, USA
e-mail: mhdurfee@mtu.edu

K. Szálkai and M. Szalai (eds.), *Theorizing Transboundary Waters in International Relations*, Springer Water, https://doi.org/10.1007/978-3-031-43376-4_9

> When the Creator was preparing to bring humans onto the earth, He called a grand council of all the animal people, plant people, and everything else. In those days, the animals and plants were more like people because they could talk. He asked each one to give a gift to the humans—a gift to help them survive, since humans were pitiful and would die without help. The first to come forward was Salmon. He gave the humans his body for food. The second to give a gift was Water. She promised to be the home to the salmon. After that, everyone else gave the humans a gift, but it was special that the first to give their gifts were Salmon and Water. When the humans finally arrived, the Creator took away the animals' power of speech and gave it to the humans. He told the humans that since the animals could no longer speak for themselves, it was a human responsibility to speak for the animals. To this day, Salmon and Water are always served first at tribal feasts to remember the story and honor the First Foods. (CRITFC 2013)

9.1 Introduction

If there were a single face of injustice in the Columbia River Basin shared by the US and Canada, it might be the Chinook salmon.[1] Hydropower utilization and environmental change have turned the Columbia and its tributaries into a mixed-use river basin with considerable industrial development in some locations. As a consequence, the number of salmonids declined sharply throughout the twentieth century. The presence of salmon is historically intertwined with the economic, social, and cultural life of Indigenous tribes within the transboundary Columbia River Basin. Within the basin's Indigenous communities, the most important aim has long been the return of salmon. Recently, the salmon issue has surfaced again due to the current renegotiation of the 1961 Columbia River Treaty, which regulates flood control and hydropower production between Canada and the United States.

The two countries are committed to continuing their extended water cooperation in the form of a renewed and modernized agreement. The official bilateral meetings began in 2018 through the usual approach of interstate water diplomacy, with the participation of negotiation teams that represent the two sovereign states of the US and Canada. The two states are seeking to preserve the good from the older treaty—power, flood control, and irrigation. They have also indicated a desire to modernize it in line with the changed circumstances, particularly related to what we know about ecosystems in the 2020s.

Improved care for ecosystems is strongly supported by many tribes in the region. However, the return of salmon, which is crucial for ecosystem function and the Indigenous peoples of the basin, is not on the original list of topics for discussion, despite the requests of tribes on both sides of the border to include salmon as a specific topic in the negotiations (CRITFC 2021a). Beyond this, the request that each negotiating team include an Indigenous person has gone unaddressed, although both sides have declared commitment to the participation of tribes on advisory committees. These decisions have evoked strong objections to denying Indigenous peoples a direct role

[1] The most well-known salmonids in the Columbia River Basin are the chinook, the coho, the sockeye, and the steelhead. Salmon are anadromous, which means that they are born in freshwater, spend most of their lives in saltwater, and migrate back to freshwaters to spawn.

in the future of the basin, a matter of direct concern to them and their governments. The US and Canada see this issue differently, as an international agreement with transnational elements in both societies.

Still, from the perspective of the national governments, the special legal authority of the tribes has been partly acknowledged by setting up a Sovereign Review Team (SRT)[2] in the US to ensure that the many sovereigns in the basin could work on the preparations of the negotiations together. In 2019, Canada initiated the observer-level participation of its three concerned Indigenous tribes on a nation-to-nation basis. The US did not follow this example, although as a response, it began to regularly invite Indigenous experts to the negotiation meetings. In that sense, sovereignty was partly acknowledged. Nevertheless, for the tribes that is not enough. As one elder put it in a 2022 Zoom meeting attended by over two hundred Americans and Canadians interested in the Indigenous views on salmon, "We are not at the table, but near it" (British Columbia 2022). In the past, Indigenous peoples had very little to do with such negotiations or in within-country decisions on dams. Thus, they continue to press for their cultural and political rights in the basin.

Consistent with normal practice in domestic policy, the American and Canadian governments have used a stakeholder approach within their broader discussions. Thus, while there is disappointment among the tribes of the basin over the absence of an Indigenous member on the formal negotiation team, they play a critical legal and expert role in the management of the basin. What makes this "normal practice" less so, is the element of international diplomacy. Within diplomacy, multi-stakeholder approaches are less common, although they have the potential to transform the status of Indigenous actors in transboundary water diplomacy and cooperation, increasing their opportunities of participation. However, the normative consequences of this transformation related to the heritage of colonialism and the respect towards Indigenous rights and sovereignty are more controversial, as the functional and interest-based approach of multi-stakeholderism may relegate more delicate, but fundamental issues to the background.

This chapter argues that more than listening to Indigenous voices is essential for the modernization of the Columbia River Treaty (CRT), and this implies a new understanding of water diplomacy compared to the negotiations of the original document. The Indigenous tribes are already embedded in the multi-stakeholder management of the basin with state and federal agencies and regional authorities. The tribes want to be represented on the negotiating teams because they are also sovereigns and because their cultures were built over the millennia in multiple relationship with the land and water of the basin. The authors of this chapter review and analyze how the approach of multi-stakeholder diplomacy in addition to normal interagency/intergovernmental relations has been transforming the position of Indigenous peoples both in the current negotiations and in the decade-long review process that preceded it. They argue that the renegotiation process needs to go beyond multi-stakeholder

[2] The Sovereign Review Team comprised the four Northwest states of Washington, Oregon, Idaho, and Montana, eleven federal agencies, and fifteen Pacific Northwest tribal governments (United States Entity 2012).

approaches and embrace the sovereignty of the Indigenous partners. The salmon issue, should it ultimately be addressed, could prove transformative for the relationship between Indigenous people and their tribal governments to other sovereigns. It would mark a significant stride towards reconciliation and decolonizing policy in the US and Canada.

One idea for resolving the salmon issue is to breach dams in Idaho to allow more salmon to make it further up the rivers of the basin. This solution has environmental, sport fishing, and Indigenous support. The dams are no longer needed for power generation, and natural resource and environmental agencies are for their removal. The governor of Idaho opposes the idea, but a US House of Representatives member (who was re-elected in 2022) advocates for it. President Biden supports the proposal, but Republicans in Congress plan to propose a law forbidding breaching dams (Mapes 2021; Stuttaford 2022; Yachnin 2022). The diplomacy for the treaty will be more than complex, and the outcome for salmon might be more of the same or a big surprise of advancing cultural and ecological well-being and justice.

9.2 Indigenous Peoples in Multi-Stakeholder Water Negotiations

With the primacy of the Westphalian state and the emphasis on state sovereignty in international law, Indigenous peoples have mostly been marginalized and omitted from direct participation in diplomatic processes (Katrandijev 2006: 123). The exclusion of people of non-European heritage from crucial decisions has been part of negotiating international treaties. Indigenous peoples did not meet the requirements of “civilized nations”, thus they could not participate in the development of international law.[3] At the same time, international law provided concepts that allowed Europeans to act in the name of a “civilizing mission” and “protection”, justifying intervention and conquest in “uncivilized” territories. In colonized areas, international law enforced the colonizer’s power through the idea of the “sovereign equality of states”, by denying the self-governing sovereignty of native peoples in important ways (Anghie 2006: 745). The US and Canada are no exception to this pattern, and the problem of the colonial exclusion of Indigenous peoples from crucial decisions has historically been present in transboundary water issues regarding the Columbia River Basin, and the negotiations about the Columbia River Treaty themselves.

The exclusive understanding of diplomacy as relations between independent, sovereign states, however, has been declining in recent decades. Although states still have primacy, numerous other actors have been gaining influence in the international community, eroding the Westphalian system and its approaches towards diplomacy. This diversification and the rising complexity of international relations through the

[3] Today the term “civilized nations” is to be avoided, and there is no distinction between “civilized” and “uncivilized” nations in the practice of international law. Decolonizing international law is, however, far from being complete (UNGA International Law Commission 2019: 52–53).

processes of globalization and localization have called for new diplomatic paradigms to adapt to the changing context.

The recently evolved multi-stakeholder diplomacy approach means "an equitable interaction between a multitude of actors (state and non-state) of varying power and position" (Katrandijev 2006: 124), which fosters finding mutually beneficial solutions for particular issues under negotiation, through the efficient use of the diverse resources of the participating stakeholders. Multi-stakeholder diplomacy values each actor's "unique perspective and expertise" (Hemmati 2000: 7). As Jansen and Kalas (2020: 2) write, cooperation based on a multi-stakeholder approach "is greater than the sum of its parts and it is about creating lasting, i.e., sustainable, and meaningful impact at all levels of action through maximizing synergies." Thus, multi-stakeholderism is also potentially able to provide the impetus for challenging established power relations (Ratner et al. 2018: 6). This format of dialogue goes well beyond the patterns of traditional interstate diplomacy and back to the pre-modern forms of diplomacy, when non-sovereign actors were also entitled to participate in diplomatic processes and sit at the negotiation table as the proactive "producers" of diplomacy (Hocking 2006: 16–17).

The complexity and multi-dimensionality of the management of transboundary natural resources means that no stakeholder, be it as potent as the state itself, is able to deal with the related issues alone. Common-pool resources need collective management (Jansen and Kalas 2020: 2) in line with collective interests, which necessitates the participation of resource-sharing stakeholders in the decision-making processes. The implementation of decisions based on one-sided interests, articulated by the most influential actor, may seem to be beneficial in the short term. However, in the long run, particularly in the case of shared natural resources, this often does not bring a real solution, and the consequences concern the entire community of stakeholders, including the initiator itself.

Regarding transboundary river treaties, state-centered diplomacy has traditionally overlooked the interests of other actors, including tribal governments or Indigenous individuals making common cause with similarly minded people and interest groups from "Settler society". The "greater interests" of the state have regularly overwritten the needs of the communities that were radically exposed to decisions taken with little thought of their impact on Indigenous communities (or often with clear thought that the change would weaken the community). In the Columbia River Basin, Indigenous peoples on both sides of the border were rarely considered. Thus, the voices of traditional knowledge and experience regarding water-related issues were discounted, silenced, or ignored. Compared to these practices, multi-stakeholder diplomacy offers inclusive, participatory solutions for involving non-state actors in negotiations about transboundary river treaties, including Indigenous peoples. In the multi-stakeholder model, crucial decisions about a river basin are more likely to be based on the carefully articulated and weighed interests of participants, which leads to less uncertainty, more legitimacy and more accountability related to their decisions. The multi-stakeholder participatory spaces of dialogue that are thus created may prevail and support the efficient implementation of the decisions, contribute to the monitoring of the related activities, offer a framework for follow-up discussions and reviews, and maintain the initiated network of cooperation, thus strengthening transboundary river treaties.

Nevertheless, multi-stakeholder water diplomacy does not offer a silver bullet for settling the issue of Indigenous participation in the decision-making processes concerning Indigenous interests. Considering Indigenous peoples stakeholders namely does not automatically mean the recognition of their rights and sovereignty, and even sitting at the negotiation table is often not enough to exert meaningful impact on the decisions (Larson et al. 2022). At the state level, the Columbia River Basin now has dozens of cooperative relations between tribal governments and other managers handling the execution of decisions made elsewhere. Sometimes the results of the tribal work changes practices among experts, but it has, as yet, not had as much impact on political decisions. At the international level the problem of not having a direct voice at the diplomatic table is further amplified by the fundamental idea of the modern state system that only one voice, that of the sovereign international state, can be binding. An observer or an Indigenous expert on salmon can provide information to the state, but unless they are on the actual negotiating team their interests might not get expressed. In general, equitable interaction among stakeholders with different rights and competences still depends on the willingness of the state to ensure their status. Power relations in multi-stakeholder negotiations favor the state with its developed and efficient toolkit of traditional diplomacy, and non-state actors may thus easily be relegated to the background (Mapendere 2008: 67–68). The state can restrict the number and capabilities of the stakeholders through its official recognition, and it also has a disproportionate influence over the priorities of the agenda. As a consequence, multi-stakeholderism has the potential to silence issues of ethics, respect, and the recognition of Indigenous rights and sovereignty in the name of functionality and shared interests.

The voice of Indigenous peoples is especially important in terms of the governance of natural resources. Based on their belief systems, Indigenous peoples often consider natural resources as related to spiritual life, which, as several case studies indicate, has significantly contributed to the conservation of these resources (Chunhabunyatip et al. 2018). Indigenous knowledge about their traditional environment is grounded in centuries of observation, coexistence, and interdependence with nature, the sacredness of which is intertwined with the survival of humans. Indigenous peoples are thus not only stewards or stakeholders of natural resources from an economic point of view, they also have a moral obligation to protect them (Quaempts et al. 2018; Samala 2021). They are connected to these resources in a cultural and social sense as well, and they are dependent on them to preserve and maintain their identity. As a consequence of this close connection, the deterioration of the environment concerns Indigenous peoples disproportionately, and they are more exposed to the harmful impact of environmental change. Being excluded from decision-making increases this exposure, prevents Indigenous peoples from preserving their traditional resources, and deprives them of essential elements of their culture and identity. Furthermore, Indigenous communities often lack access to the allocation of financial resources, which hinders them in protecting their traditional environment (Garneau 2022). Thus, their equitable participation in decision-making processes is not only indispensable from the perspective of Indigenous rights and sovereignty, it is also of key importance for the conservation of natural resources for all of society.

Multi-stakeholder practices emerged within the Indigenous communities of the Columbia River Basin before the appearance of multi-stakeholder approaches in international diplomacy and before other fields had opened participatory spaces for Indigenous peoples. For example, the Columbia River Intertribal Fisheries Commission (CRITFC) was formed in 1977 to protect their 1855 treaty rights to fish. The Canadian Columbia River Intertribal Fisheries Commission (CCRIFC) came later, during the 1990s, to facilitate coordination with the US. In the 1980s and 1990s, governments in other countries also began to acknowledge that Indigenous peoples were not mere beneficiaries of government programs but active, informed partners with their own interests and agency (Jansen and Kalas 2020: 2). This trend was strengthened in 1989 by the ILO's Indigenous and Tribal Peoples Convention (No. 169), which states that Indigenous peoples have "the right to decide their own priorities for the process of development as it affects their lives, beliefs, institutions and spiritual well-being and the lands they occupy or otherwise use, and to exercise control, to the extent possible, over their own economic, social and cultural development. In addition, they shall participate in the formulation, implementation and evaluation of plans and programmes for national and regional development which may affect them directly" (Article 7). In 2007, the United Nations Declaration on the Rights of Indigenous Peoples (UNDRIP) confirmed the right of Indigenous peoples to participate in the decision-making affecting their rights (Article 18), and it established the principle of free, prior and informed consent (Article 19).

The US and Canada did not ratify the ILO Convention, and they first voted against the UN Declaration, emphasizing their reservations. However, both states have since given their support to UNDRIP, recognizing the international legal commitment to the equitable participation of Indigenous peoples in the affairs that concern them. The Organization of American States (OAS), an international organization founded in 1947 to promote cooperation between the nations of the Americas, issued a regional document, the American Declaration on the Rights of Indigenous Peoples, which also confirms this principle.

Overall, it is widely accepted that multi-stakeholder initiatives have a positive impact both on articulating and representing Indigenous interests and their involvement in participatory frameworks of decision-making and coordination (Rodriguez and Sarmiento Barletti 2021; Sarmiento Barletti et al. 2022). Critics of multi-stakeholderism, however, doubt the empowering potential of this approach in the case of Indigenous peoples (Martens 2007; Larson et al. 2022). The most important concern in this respect is the tendency of multi-stakeholderism to prefer interests to values, and the state to non-state actors. This questions whether multi-stakeholderism can account for the recognition of Indigenous rights and sovereignty, and whether it can provide meaningful relief for their historical marginalization in decision-making processes. As Cohen and Norman (2018: 8) argue about Indigenous rights, although they give "many Indigenous communities a 'seat at the table,' the 'table' needs to be rebuilt," to include Indigenous perspectives and account for their historical rights from the start. According to this standpoint, a structural change is needed beyond the "add Indigenous peoples and stir" approach, recognizing the inherent rights and sovereignty of these peoples.

9.3 Indigenous Tribes and Salmon in the Columbia River Basin

The Columbia rises in the Rocky Mountains of Canada, and flows into the Pacific Ocean at Astoria, Oregon, about 50 miles from Portland. It is the fifteenth longest river in North America with the sixth largest volume of runoff. On rare occasions during very wet years, it delivers more water to the Pacific than the Mississippi does to the Gulf of Mexico. The Columbia basin is about the size of France and has over 40 tributaries, with the Kootenai, the Flathead/Pend Oreille, the Snake and the Willamette Rivers being the largest. The basin drains parts of British Columbia and Alberta in Canada (102,300 square kilometres) and parts of Washington, Oregon, Idaho, Montana, and small sections of Nevada, Utah, and Wyoming in the US (567,000 square kilometres). About 25% of the river flow originates from Canada. The seasonal variation of the flow is very high, which makes the basin prone to floods (Bonneville Power Administration 2001: 4–5).

In the US-governed part of the Columbia River Basin, fifteen officially recognized Tribal Nations have territories, authorities, and responsibilities that are affected by the Columbia River Treaty.[4] On the Canadian side of the border, three First Nations are recognized in the basin: the Ktunaxa, the Okanagan, and the Secwepemc. Tribes and First Nations possess considerable autonomous agency in the Columbia River Basin as government entities and as once fully self-governing polities. The tribes are sovereign in important ways, although in reality they hardly have the authority that the Indigenous treaties suggest they should have—for example, in terms of fishing rights (Thompson et al. 1996; Walter et al. 2000; Bell 2015). Both the US and Canada maintain a government-to-government relation with their officially recognized Indigenous peoples, which is grounded in their constitutions. Indigenous peoples are not treated as interest groups or non-governmental organizations, they are sovereigns under their own governance structures, with the right of tribal self-determination (US Department of Justice 2022; Department of Justice Canada 2021). In US and Canadian law, "Indian law" is a distinct, although quite incoherent body of law, somewhere between international law and federal law (Mason 1983; Deloria and Wilkins 1999). Treaties and the ensuing laws to give them effect assign fishing rights to the tribes

[4] They are the Burns Paiute Tribe, Coeur d'Alene Tribe, Confederated Salish & Kootenai Tribes of the Flathead Nation, Confederated Tribes and Bands of the Yakama Nation, Confederated Tribes of the Colville Reservation, Confederated Tribes of the Umatilla Reservation, Confederated Tribes of the Warm Springs Reservation of Oregon, Cowlitz Indian Tribe, Fort McDermitt Paiute Shoshone Tribes, Kalispel Tribe of Indians, Kootenai Tribe of Idaho, Nez Perce Tribe, Shoshone Paiute Tribe of the Duck Valley Indian Reservation, Shoshone-Bannock Tribes of the Fort Hall Reservation, and the Spokane Tribe of Indians.

of the Columbia. Nevertheless, it is a constant battle to secure them and to press for better protection of fish, especially the salmonids.

Originally, salmon runs were estimated to be 10–16 million fish per year in the basin. After a sharp decline in the twentieth century, their number reached a historical low of 749,000 in 1994 (Lichatowich et al. 2017: 3), while now it is in the 1,000,000–1,200,000 range (Harrison 2021), less than one tenth of the original number of runs, with the majority of fish (approx. 75%) raised in hatcheries (Bell 2015: 270). Among the sixteen interior salmonid species, four are extinct, seven are listed as threatened or endangered under the Endangered Species Act (ESA), while the remaining five comprise only a fraction of their historic numbers (NOAA Fisheries 2022). In sum, the number of salmon runs has not increased considerably in the past few decades, in spite of the various restoration programs and ongoing initiatives to bring salmon back – like the initiative of the Northwest Power Planning Council,which declared the aim to reach at least 5 million fish annually more than 40 years ago (Lichatowich et al. 2017: 5).

The first large structure changing the flow of the river was the Rock Island Dam, which was built in 1894 and strengthened in 1932 (Cohen and Norman 2018: 5). From the aspect of the salmon and Indigenous peoples, the building of the Grand Coulee Dam, which opened 1942, brought the most radical change, ending most of the salmon runs on the upper part of the Columbia. The impact of the Grand Coulee Dam eliminated the second most important salmon fishery at Kettle Falls (US Bureau of Reclamation 2021). The most important one, Celilo Falls had served as a fishing spot and community gathering site for more than 15,000 years. It was inundated in 1957, with the construction of the Dalles Dam. The Columbia River Treaty in 1964 opened further opportunities for building dams. Reservoirs built on the upper parts of the river flooded culturally important areas belonging to Indigenous peoples in the territory of Canada as well (Cohen and Norman 2018: 15). Today there are more than 480 dams in the river basin, more than 30 of which are large, and several are impassable for fish (Columbia Basin Trust 2018; Northwest Power and Conservation Council 2022a).

Salmon were not just a way to provide food and trade in the Columbia Basin. Salmon has been thoroughly intertwined with Indigenous peoples' spirituality, their belief systems, the organization of work, social and cultural life, diets, and everyday activities. As Rodney Cawston, Chairman of the Colville Tribal Business Council says, "Our culture, our religion, our way of life. Everything revolved around having salmon come up the river" (Collins 2019). Humans are traditionally called *Wy-Kan-Ush-Pum* or Salmon People (CRITFC 2021b), and tribe representatives often refer back to the legend that "the salmon was put here by the Creator for our use as a part of the cycle of life (...) Their returning meant our continuance was assured because the salmon gave up their lives for us" (HighEagle, quoted in CRITFC 2021b). With the restricted salmon runs, numerous elements of traditional Indigenous cultures disappeared, including customs, ceremonies, community networks, and the teachings of the elders, or practices like the special leading role of women in the trade of dried and powdered salmon. Dams and reservoirs also destroyed numerous cultural, historic, and spiritual sites (Ween and Colombi 2013: 482–483).

Beyond salmon, the environmental change fostered by the Columbia River Treaty has had a transformative impact on the whole ecosystem of the basin. As Indigenous peoples traditionally exist in a closer relationship with nature, the consequences have a disproportionately negative impact on their living circumstances. The Indigenous peoples of the United States have created three alliances in the Columbia River Basin in order to represent their joint interests related to their traditional ecosystem: the Columbia River Inter-Tribal Fish Commission (CRITFC), the Upper Columbia United Tribes (UCUT), and the Upper Snake River Tribes Foundation (USRT). In Canada, the Canadian Columbia River Inter-Tribal Fisheries Commission (CCRIFC) fills a similar role. The issue of the ecosystem is of a great importance for each of these alliances, and the reintroduction of salmon is in the focus of a significant part of their activities. They are also important in maintaining the cultural heritage related to the river and its natural resources, and sharing elders' stories about the customs and traditions related to the reciprocal relationship between the community and nature is an essential part of their work (Cohen and Norman 2018: 15–16).

Throughout the centuries, the Indigenous peoples of the US and Canada have litigated their right to fish, and more recently they have built extensive expertise that non-Indigenous scientists recognize and value (Weaver 1997; CRITFC 2021a). A vast corpus of case law protects the rights of Indigenous peoples to their territories, as well as their traditional rights concerning tribal fisheries, confirming them in their status of co-managers of natural resources (Cohen and Norman 2018: 17). However, experts and lawyers have not been able to change the unexpected consequences of the environmental change that was generated through the industrial development of the basin, especially by building dams on the Columbia. The environmental change fostered by the Columbia River Treaty transformed the everyday lives, cultures, and even identities of Indigenous peoples, who traditionally considered themselves stewards of their environment, with "a very clear mandate to protect the land and the resources" (Matthew, quoted in Schafer 2022). However, they were silenced when the most important decisions were made about them.

9.4 The Original Negotiations of the Columbia River Treaty

The Columbia River Treaty is often considered a "model of international cooperation on transboundary water management and hydropower development" (e.g. Government of Canada 2022). At the international level, US–Canada transboundary water relations in general comprise a special case, notable for several exceptional factors. The two countries have numerous shared water bodies, and the border between them takes advantage of rivers and lakes wherever possible. A great power and a middle power are involved in the long history of cooperation between the two states, both of which are highly developed countries with many shared cultural elements. Thus domination is unlikely, and theories of hegemony do not quite explain their relations. They share a history of treaties through which they have repeatedly solved transboundary water issues with the help of negotiations.

The negotiations about the Columbia River Treaty began as a consequence of two emerging issues in the 1940s: the increasing number of problems caused by floods, and the rising energy demand of the rapidly growing population in the Pacific Northwest region. The process took almost two decades. The International Joint Commission (IJC) of the US and Canada (established in 1909 with the Boundary Waters Treaty) created the International Columbia River Engineering Board to support the negotiations with technical expertise, looking into the possibilities of a joint development of the Columbia basin. Their final study (International Columbia River Engineering Board 1959) was published after fifteen years of work, suggesting the construction of new dams and reservoirs in the upstream parts of the Columbia and its tributaries, and making recommendations on sharing the benefits of the joint development of the basin. The direct negotiations, based on the IJC's suggestions, began in 1960 and proceeded without significant disagreement: the treaty was signed in 1961, while implementation began in 1964 (Sewell 1964; US Army Corps of Engineers and Bonneville Power Administration 2009). As the time and the context of the original treaty would suggest, Indigenous peoples did not participate either in the negotiations, or in the preparation process—their rights guaranteed by the original treaties with the colonizers were not respected, their knowledge and their technical expertise were completely ignored, as were their economic, social, and cultural needs.

There is no termination date in the agreement. The US and Canada[5] made a long-term commitment to the treaty: only after 60 years are they to decide if they wanted to terminate the agreement, renegotiate it, or leave it operating according to the original decisions. This date will come in 2024. Even if a party announces a desire to negotiate, the treaty provides a waiting period of ten years, which slows down change and allows public binational discussions and studies before the negotiations (Article XIX). The CRT focuses solely on the issues of flood control and hydroelectric development between the US and Canada, for the implementation and governance of which it created two national organizations: the US Entity and the Canadian Entity (Article XIV). The US Entity is comprised of the representatives of the Bonneville Power Administration and the US Army Corps of Engineers, while the Canadian Entity is the British Columbia Hydro and Power Authority. The CRT also set up a joint body of experts, the Permanent Engineering Board (PEB) to assemble reports and conduct inspections. PEB consists of four members, two appointed by the US, and two by Canada (Article XV). PEB also has the competence to assist in settling disagreements between the Entities, although in this case, the CRT refers to the IJC as the primary tool for arbitration (Article XVI).

Under the 1909 Boundary Waters Treaty (BWT), the IJC manages transboundary water issues at the request of the two national governments. Since 1941, it has had an International Columbia River Board of Control, which consists of a Canadian

[5] Although the Entities govern the CRT, termination and renegotiation are the competences of the two governments. In Canada, the Province of British Columbia maintains most rights, benefits, and obligations related to the CRT based on the Canada-British Columbia Agreement (1963). However, based on the constitution, the negotiation of international treaties belongs to the government of Canada, while British Columbia's consent is required for any modifications.

representative from the Ministry of Environment and Climate Change and an American representative from the US Geological Survey. Its purpose is to monitor and recommend any actions necessary due to the changes caused by the Grand Coulee Dam (IJC 2022). The BWT also contains a general arbitration clause that allows the US and Canada to submit any dispute to them. Although the IJC has never been asked to arbitrate between the US and Canada, it has the potential to do so, and it does use extensive expertise to manage conflicts over use. Indigenous peoples do not appear in the text of the CRT, either as actors with specific interests and rights, or as mere subjects of the regulations. There is no mentioning of any kind of Indigenous representation in the established bodies, nor of a need to consult or share information with Indigenous peoples.

The CRT required the building of four treaty dams, three in Canada (Duncan 1968; Hugh Keenleyside/Arrow 1969; Mica 1973), and one in the US (Libby Dam 1973), to increase storage capacity and electricity creation. Their development facilitated the further utilization of the Columbia for hydropower use, increasing the number of dams and reservoirs. These events led to a previously unseen level of environmental change in the basin, inundating hundreds of thousands of hectares and displacing thousands of people (British Columbia 2023a). This has affected the cultural and natural resources of Indigenous peoples in a harmful way, making them suffer disproportionate disadvantages without any influence on the relevant decisions.

The treaty also regulated the benefits of the joint development of the basin. The regulations are based on the principle of equality in the field of both flood control and hydropower production. In return for flood control, Canada is entitled to 50% of the estimated value of the prevented damages in the US This commitment was paid in lump sum payments, reaching USD 64.4 million until 1973, which is considered to be sufficient until 2024. In terms of hydropower, the treaty guarantees 50% of the estimated downstream power benefits to Canada. This was sold to US utilities between 1973 and 2003 for USD 254 million; since then, it is delivered on a daily basis to Canada (US Army Corps of Engineers, Bonneville Power Administration 2009: 6). This arrangement is the Canadian Entitlement, which is a very strong driver of the renegotiation process today.

The US argues that it needs reforms, because Canada's share should be based on the value of their flood and power contribution based on *coordinated* operations. The US argues that the payments should reflect the great value the American contribution brings to the system for Canada, while Canada claims that they are the ones who receive disproportionately fewer benefits from the cooperation. The amounts of water traded between the two parties may need to be altered, as do the funds sent to Canada for providing water storage for flood control (Reimer 2021; Stern 2022). The negotiating parties often emphasize that the first priority of the modernization of the treaty is the clarification of the conditions regarding the continuation of the original regulations about flood control and hydropower production, especially in the case of the Canadian Entitlement, without which Canada is not likely to continue its treaty obligations (Stern 2022:1, 9). "Additional interests" (US Army Corps of Engineers, Bonneville Power Administration 2009: 8), including those of Indigenous peoples concerning the ecosystem and the reintroduction of salmon, are often relegated to the background.

Although the Columbia River Treaty is often referred to as "the standard against which other international water coordination agreements are compared" (US Army Corps of Engineers, Bonneville Power Administration 2009: 7), it is clear that it was negotiated in a context that cannot be applied as a guiding framework today. The fact that 60 years ago the negotiations were carried out in the traditional sense of interstate water diplomacy, silencing the voices of Indigenous peoples and their rightful ownership over numerous concerned territories, and ignoring the needs of the ecosystem, makes the current version of the treaty anachronistic and unsustainable in its present form. The renegotiation of the CRT raises the unquestionable need to modernize with the participation of the Indigenous peoples, respecting their inherent rights and sovereignty, and with a regard for the ecosystem, including salmon reintroduction.

9.5 Indigenous Tribes at the CRT Renegotiation Table

The US and Canada committed themselves to the renegotiation of the treaty in 2016 and 2014, respectively. By February 1, 2023, fifteen official rounds of negotiations had been conducted, the last one between January 25–26, 2023, in Vancouver, B.C. These negotiations were accompanied by several informal and technical meetings, and further proposals were exchanged (British Columbia 2023b). The process had been preceded by a multidimensional binational review (2014/2024 Columbia River Treaty Review) led by the US and the Canadian Entities. Instead of leaning on the intergovernmental IJC's recommendations, as in the case of the original treaty, this review aimed to comprise as many stakeholders' opinion as possible. This is certainly a great change compared to the negotiations of the original treaty for the various stakeholders, including Indigenous peoples (United States Entity 2012). The negotiations make extensive use of the outcomes of the 2014/2024 Columbia River Treaty Review based on extended consultations with stakeholders and technical experts, which produced two recommendations for state principles (US Regional Recommendations and B.C. Decision) and a series of studies, while fostering regional discussion about the modernization of the treaty. However, the negotiations themselves still began in the spirit of traditional diplomacy in 2018. The first meetings were organized with the participation of the official representatives of the two sovereign states of the US and Canada, behind closed doors, without the presence of Indigenous peoples. As the meetings are not public, and press releases are rather laconic, the two recommendations are still the most detailed documents available on the standpoints of the US and Canada.

Neither the US, nor Canada denies the need to modernize the treaty with the participation of Indigenous peoples, and both negotiating states are committed to the principles of multi-stakeholderism, Indigenous rights, and environmental sustainability. In terms of multi-stakeholderism, the two countries use integrated international water management at the administrative level. As a consequence, the basin boasts extensive networks of experts, such as scientists, engineers, economists, and lawyers, including scientists and lawyers from the tribal governments. The experts of the tribes use the language of modern science while also being informed by their culture. In addition, the tribes have excellent lawyers, who push wherever possible to promote and protect Indigenous rights and sovereignty. Meanwhile, citizens, firms, interest groups, and non-profit organizations participate in decisions through numerous committees, advisory boards, and various other forms, fostering multi-stakeholderism within states and, in some cases, over the borders as well.

As noted earlier, the tribes pressed for breaching some dams to bring the salmon back all the way to Canada. Trout Unlimited, a fishing/conservation NGO, joined the chorus and called for breaching some of the dams on the lower Snake River. Republican Congressman Mike Simpson from Idaho agreed and proposed a 33.5-billion USD fund to do so, and to provide for other sustainable forms of energy and habitat restoration (Mapes 2021). Later, the Nez Perce Tribe announced that the National Congress of American Indians had passed a resolution supporting Representative Simpson's plan (Ramey 2021). The Idaho Conservation League (2021) also supports Simpson, and calls to breach the dams on the lower Snake also showed up on Facebook. In July 2022, NOAA published a draft report titled "Rebuilding Interior Columbia Basin Salmon and Steelhead," with the contribution of the Nez Perce Tribe. The draft claimed that breaching the four Snake River dams in question is "essential", with support for its conclusions from the Biden administration. (The final report was published in September as NOAA Fisheries 2022). Once the report came out, opposition grew from the governor of Idaho, business groups, and many members of the Idaho legislature who worry that they will lose a significant source of their electricity and that breaching dams would reduce a "green" source of energy—hydropower—compared to oil and gas (a problem that does not seem to be a problem at all, given improvements in energy technology and management) (Stuttaford 2022; Yachnin 2022). The idea of breaching dams threatens existing water cooperation by redistributing the benefits of the river among the various players. The divide over breaching dams on the lower Snake River also illustrates how difficult it can be to address a problem of justice related to Indigenous rights due to a tight link between the domestic and the international level.

As for Indigenous rights, public opinion is mixed but slowly changing, partly due to the proactive efforts of the Native American community (e.g. the Reclaiming Native Truth Project by the First Nations Development Institute and Echo Hawk Consulting [2023]). In the US, the Obama administration declared its support for the United Nations Declaration on the Rights of Indigenous Peoples in 2011, and stated that "U.S. Government efforts to strengthen the government-to-government relationship with tribes cannot be limited to enhancing tribal self-determination" (U.S. Department of State 2011). The November 2020 election of Democratic President Joe Biden has been considered a positive development for Indigenous rights, which suggests that tribal voices will receive more attention in the near future. "There is reason for optimism" according to a policy analyst from CRITFC: "The view of the tribes will be taken much more seriously under this administration" (Pynn 2021). However, unlike in Canada, there is no systematic and centrally coordinated initiative to review the harm Indigenous communities suffered as a consequence of colonization, and there are also no official attempts of reconciliation.

In Canada, the recent increase in the awareness of Indigenous rights can be traced back to the functioning of the Truth and Reconciliation Commission between 2007 and 2015. The primary task of the Commission was to investigate the legacy of the Indian Residential Schools system, which separated 150,000 Indigenous children from their families and cultures exposing them to various abuses, and to help reconciliation. In its closing executive summary (Truth and Reconciliation Commission of Canada 2015), the Commission identified 94 "calls to action" for reconciliation and a renewal of relationship in a much broader sense between Indigenous peoples and all Canadians. In 2016, the government expressed its unqualified support for the United Nations Declaration on the Rights of Indigenous Peoples, and committed itself to implementing it in a transformative way (Fontaine 2016). The extensive discourse on Indigenous rights, a positive affirmation of which is widely present in official state and provincial documents, is also driving attention to the role of Indigenous peoples in the renegotiation of the Columbia River Treaty. Nevertheless, this does not mean that each principles are applied in practice as well, for example, a dam is being built on the Saaghii Naachii/Peace River in British Columbia against the strong objections from Tribal governments (Behn and Bakker 2019).

As for environmental sustainability, over the decades, the scientific understanding of the ecosystems of the basin has shown that the loss of fish has had a significant impact on the overall health of the basin. Thus, the physical structure of some of the dams has been adjusted in the hope of returning the salmon and other fish. Sometimes the managers of the dams literally truck fish to get around dams that have no fish ladders (Northwest Power and Conservation Council 2022b). Both sides of the border have fish hatcheries. The Indigenous fisheries are crucial players in the effort of bring salmons back to the rivers of the basin, which is not only important for cultural reasons. There is a rising awareness of the importance of salmon for the whole ecosystem of the river, significantly fostered by Indigenous experts and organizations. The great salmon runs of the past brought important nutrients from the ocean to the land and water. The fish fed people, bears and other creatures of the land and the river (Rahr 2016).

Meanwhile, the original driving forces that initiated the focus of the Columbia River Treaty on the economic development of the basin, i.e. jobs during the depression, nuclear weapons during WWII and the Cold War, hydropower, flood control, better maritime transportation, and recreation, have changed. The economics of power generation and demand are different as well. There is no need for more dams, as illustrated by the initiatives to breach some of them. Power demand in the basin can be met due to technological innovations in highly efficient turbines and more efficient power technology in homes as well as in hydropower generation. Demand for energy has also been decreasing with the decline of industrial production in the basin, since for example a previously much-desired aluminium plant in the State of Washington has closed (Brunnel 2017). Flood control, of course, has not lost its relevance, massive floods indeed hit the Nooksack River in 2022, flooding parts of British Columbia and Washington state (Cotsirilos 2022). Better ecosystem functioning could moderate the effects of both.

The impact of global warming is visibly present in the Columbia River Basin. Climate change eats at the very headwaters of the river system, as the glaciers in the Rocky Mountains are slowly melting. The weather is also changing, and precipitation is more variable. The water in the rivers is getting warmer, and thus an already compromised ecosystem is expected to go into further stress. In September 2022, the US declared fishery disasters due to the failure of the salmon run in the US Pacific Northwest (Blakemore 2022), which had been triggered by the unusually warm waters due to severe drought and high temperatures. The integration of the ecosystem function as the third leg of the Columbia River Treaty would contribute to the improvement of the riverine habitat. A healthier river would also run cooler and is likely to moderate some of the effects of climate change.

Overall, in the multi-stakeholder approach focusing on common interests and synergies, the present trends not only support, but also necessitate the inclusion of Indigenous peoples and their preferences in the negotiations. The increasing awareness of Indigenous rights in both negotiating states provides a favorable context for their participation, while under the above-discussed circumstances, their traditional knowledge and expertise are invaluable for solving the most pressing issues related to the ecosystem of the Columbia basin. Related to these issues, the renegotiation process of the Columbia River Treaty has to provide an answer to three interrelated questions concerning the interests, rights, and sovereignty of Indigenous peoples: (1) Will the ecosystem function be included as the third leg of the CRT? (2) Will the issue of salmon be included explicitly in the CRT? (3) How can Indigenous peoples let their voice be heard in the negotiation and the decision-making process? In the following sections, we review the participation of Indigenous peoples in the renegotiation process in both states, and evaluate the answers to these three questions.

9.5.1 The Case of the US

The Indigenous peoples, aware of the potential impact of the renegotiation of the treaty, established the Coalition of Columbia Basin Tribes in 2009, comprising the fifteen US tribes in 2009. The aim of the Coalition is to speak with one voice, and present "a common vision of restoring ecosystem function and resiliency to the Basin" (Letter of the Columbia Basin Tribes Coalition 2014). In 2010, they issued the document Common Views on the Future of the Columbia River Treaty, expressing their criticism towards the original treaty, and emphasizing the importance of the representation of their interests "in the implementation and reconsideration" of the new treaty. Their claims in the related processes include:

- "Respect for the sovereignty of each tribal government—each tribe has a voice in governance and implementation of the Columbia River Treaty.

- Tribal cultural and natural resources must be included in river management to protect and promote ecological processes—healthy and useable fish, wildlife, and plant communities.
- Integrate the tribes' expertise of cultural and natural resources in river management.
- Equitable benefits to each Tribe in priority to other sovereign parties in Columbia River management.
- Respecting and preserving the benefits of settlement agreements with tribes.
- Recognize tribal flood control benefits.
- Protecting tribal reserved rights to current and future beneficial uses, in a manner consistent with ecosystem-based management" (Columbia Basin Tribes Coalition 2010).

Although salmon are not mentioned explicitly, the most important practical aim for the members of the Coalition is the restoration of salmon runs, especially over the Chief Joseph and the Grand Coulee Dams.

During the Columbia River Treaty 2014/2024 Review in the US, the voices of Indigenous peoples played an emphasized role. Representatives of the fifteen tribal governments participated in the Sovereign Review Team appointed by the US Entity in 2010, with representatives of four Northwest states and eleven Northwest federal agencies, while other stakeholders had more limited opportunities to express their ideas. Beyond the tribal governments, the Indigenous alliances also participated in the preparation of the final document of the review. The Regional Recommendation (2013) officially serves as an important basis for articulating the standpoints of the State Department's Negotiating Team.

The Recommendation was a reasonably big success for the representation of Indigenous interests. Although it strengthens the commitment to the continuation of the original functions of flood control and hydropower production, the text contains several expressions referring to a "healthy river system" and "environmental sustainability", and it highlights the "opportunity for inclusion of certain additional ecosystem operations to expand, enhance, and complement" the treaty. The "ecosystem-based function", strongly supported by Indigenous peoples, is presented as equal to the two original purposes. As for salmon, the Recommendation raises the issue of investigating opportunities for "restored fish passage and reintroduction of anadromous fish" together with Canada. The Coalition cooperated with Canadian Indigenous peoples to publish a plan on fish reintroduction into the US and Canadian Upper Columbia River (The Columbia Basin Tribes & First Nations 2015) to support further discussion based on the Recommendation. In terms of Indigenous representation, the text highlights the importance of "Indian treaties and tribal rights, [and] the US Government's trust responsibility to the tribes" among the responsibilities of the U.S related to the new treaty, and it emphasizes the role of "federally recognized tribes" among other stakeholders in the preparation of the Recommendation and in addressing domestic matters related to it after 2013. Following a review, the State Department accepted the document in 2016 and made the official decision to negotiate about the modernization of the treaty.

After these promising initial developments, the exclusion of Indigenous peoples from the negotiations that began in 2018 came as a surprise. The negotiating team led by Jill Smail included the representatives of the two Entity authorities: the Bonneville Power Administration and the US Army Corps of Engineers, Northwestern Division the Department of the Interior, and the National Oceanic and Atmospheric Administration (NOAA) (U.S. Department of State 2018). As a response to the dissatisfaction with this decision, the State Department stated that it "values the Tribes' expertise and experience, and is consulting with the Tribes throughout the negotiating process" (Collins 2019), and that "the best way to balance the United States' objectives and conclude a successful agreement with Canada in a timely manner is to limit the negotiating team to federal agencies" (Scruggs 2019). In spring 2019, when Canada changed its similar initial decision, and announced the involvement of Indigenous representatives in the negotiation team, the reaction of the State Department was the following: "We will continue to engage the Tribes on a regular basis as negotiations proceed," but "we have no plans to change the general composition of the team" (Collins 2019) Nevertheless, in the next round of meetings, in autumn 2019, Indigenous experts joined the negotiations on the side of the US as technical advisers. Since then, three Indigenous representatives from different tribes are regularly present in this role (B.C. Columbia 16 Sept 2019; Pynn 2021).

9.5.2 *The Case of Canada*

The three Indigenous peoples of British Columbia represent a shared standpoint regarding the Columbia River Treaty, which shares similar aims with the US Coalition of tribes. Their cooperation is fostered in the framework of the Canadian Columbia River Inter-Tribal Fisheries Commission (CCRIFC), which was already established in the early 1990s to protect fish and aquatic ecosystems, as well as represent the rights of Indigenous peoples connected to them. They treat the issue of restoring salmon populations as a "cultural imperative", intend to mitigate the harmful impact of dams and reservoirs in the Columbia Basin, and provide technical expertise (CCRIFC 2016). The cooperation of British Columbian Indigenous peoples in these fields is, however, less institutionalized than on the other side of the border, especially since the Syilx Okanagan Nation withdrew from CCRIFC in 2002. In the meantime, the visibility and awareness of Indigenous issues is greater in Canada, both in the government and in the society, due to the extended Truth and Reconciliation process, which contributes to making the voices of the three Columbia River Basin Nations heard in the renegotiation of the CRT.

The review of the Columbia River Treaty in Canada began in November 2011. According to the Public Consultation Report, the Province of British Columbia put great emphasis on learning the interests, concerns, and standpoints of Indigenous peoples and consulted with them separately, on a government-to-government basis. The report refers to the historical grievances of Indigenous peoples as a consequence of the treaty, and it also underlines the constitutional duty to recognize the rights of Indigenous peoples and the obligation to involve them in decisions that concern them (British Columbia 2014). The outcomes of the consultations are, however, not public, and since B.C. tribes did not issue a joint public document on their standpoint, we know significantly less about their self-articulated interests and opinions than in the case of their US counterparts.

The B.C. Decision itself is more reserved than the US Regional Recommendations in some aspects that hold special importance for Indigenous peoples. Both in terms of the ecosystem and in the salmon question, it refers to possible "available mechanisms outside the Treaty", claiming that "the management of anadromous salmon populations is the responsibility of the Government of Canada". The B.C. Decision explicitly argues that the restoration of salmon is not an international issue, thus, not a treaty issue, and each country is responsible for its activities in this field. Indigenous peoples do not protest publicly against this standpoint, and they are open to cooperating with the governments of B.C. and Canada separately on state-level action within the frameworks of the Columbia River Salmon Reintroduction Initiative (CRSRI 2023). A Letter of Agreement (2019) on this initiative was signed among the five parties in July 2019, on a government-to-government basis. Overall, the B.C. Decision accepts the importance of the ecosystem-based function beside the two original aims of the treaty, commits itself to exploring ecosystem-related improvements, and recognizes the need to include the aim of adapting to climate change in the new treaty, thus including the most important Indigenous priority, similarly to the US position.

Based on the B.C. Decision and the extended social discourse on reconciliation, the public did not expect that Indigenous peoples would be excluded from the negotiations. When it became obvious that they were not invited to the table, the disappointment led to a lot of public criticism against the decision. Dissent was also driven by the fact that the declaration was issued rather late before the first meeting, and the government was accused of bowing to US pressure (Mclash 2018). Grand Chief Stewart Phillip, the President of the Union of B.C. Indian Chiefs and Chair of the Syilx Okanagan Nation, referred to the decision as "an act of absolute treachery", "a fundamental betrayal", and he accused Canada of following Donald Trump's "aggressive destructive behaviour", which "undermines recognition and threatens the reconciliation path" (Crawley 2018).

In April 2019, Canada declared that the representatives of its three Columbian Indigenous peoples, the Ktunaxa, Okanagan, and Secwepemc Nations are invited as official observers to participate in the negotiations (Global Affairs Canada 2019). The three nations welcomed the decision, and since then, they have represented themselves at all the meetings. Their participation is regulated by the CRT Negotiations Framework Agreement (2019) between Canada, British Columbia, and the three Indigenous actors on a nation-to-nation basis. According to Katrine Conroy, Minister Responsible for the Columbia River Treaty, "this level of participation by Indigenous Nations in an international negotiation is unprecedented", which underlines Canada's commitment to the process of reconciliation, and contributes to making the modernization of the new treaty more effective (Conroy 2022). The Indigenous leaders, on the other hand, have stated that they "look forward to helping to ensure any new Treaty addresses the mistakes of the past," and greeted the "small but meaningful steps together on the road to reconciliation" (Global Affairs Canada 2019). The three Indigenous peoples cooperate closely on the negotiation team in the spirit of the so-called Tri-Nation Approach, representing the importance of the ecosystem function, Indigenous rights and sovereignty, salmon reintroduction, and the principle of free, prior, and informed consent (Syilx Okanagan Nation Alliance 2022; Columbia Valley Pioneer 2019).

9.5.3 Case Summary

During the renegotiation of the Columbia River Treaty, the practices of multi-stakeholder diplomacy are defining. Officially, Indigenous peoples are not 'simple' stakeholders, they have inherent rights as sovereigns over their lands. Canada places great emphasis on this and intends to make Indigenous peoples' participation in the negotiations as visible as possible. Even if the US is more reserved in this sense, the Canadian example has led to the inclusion of experts on their side as well. It is also important to note that during the preparation procedure, both states followed similar procedures of listening to the voice of Indigenous peoples and involved them in the consultations in a special position among stakeholders. Overall, both states recognize the need to include Indigenous peoples in the various stages of the renegotiation,—especially in terms of expertise, and to honor and confirm their traditional rights over the concerned territories. Thus, in the normative sense, multi-stakeholder approaches indeed contribute to the participation of Indigenous peoples in the modernization of the Columbia River Treaty, and they significantly improve their status compared to the original negotiations 60 years ago.

At the same time, there are different normative directions as well. Does the improved status mean a recognition of sovereignty with increasing influence? Indigenous peoples from both sides of the border are now present in the negotiation room, and their most important priority, adding the ecosystem-based function as the 'third leg' of the treaty seems to have been successfully addressed in principle. However, the above analysis shows that (1) Indigenous representatives are present in the government-assigned roles of experts or observers, still far from the table of decision-makers. (2) Their influence on the agenda is limited. Even if the issue of salmon was reportedly raised during the consultations several times, the Canadian government has refused to address it in the framework of the CRT. (3) There is no discourse on either side of the border about Indigenous peoples' participation in the *governance* of the new treaty.

As the interest-based multi-stakeholder approach has the potential to relegate the recognition of Indigenous rights and sovereignty to the background, it may hinder a very important additional function that is often associated with the modernizing of the Columbia River Treaty. Several Indigenous leaders (Letter of the Columbia Basin Tribes Coalition 2014) and non-Indigenous experts (Save Our Wild Salmon 2023) namely emphasize that the renegotiation of the treaty offers a historical chance to "right many wrongs"—to salmon, to the whole river basin ecosystem, and to the region's Indigenous peoples. If all three wrongs are addressed, it will be a significant act of reconciliation with Indigenous peoples and thus an important building block of decolonizing policy in the US and Canada. If this aspect is not considered explicitly, this golden opportunity will be lost.

9.6 Conclusion

The negotiation of the original Columbia River Treaty has a long history, one dating back to at least the Oregon Treaty (1846), when Britain and the US agreed on a border (without consultation with the tribes). Canada and the United States consolidated their powers over a river basin and its resources through the means of traditional diplomacy and the arrival of settlers from the North American East. The process took a heavy toll on the ecosystem and its Indigenous peoples. If the parties indeed want to modernize the treaty, the most important question about the renegotiation is not formulated in terms of interstate relations, but regarding a change of perspective.

"Being heard is half the battle" (Collins 2019), but for Indigenous peoples and their governments it is a battle Americans and Canadians need them to win. Multi-stakeholder diplomacy can indeed contribute to making Indigenous peoples' voices heard in issues related to transboundary waters. Nevertheless, this does not mean that the normative implications of multi-stakeholder approaches point to one direction. Multi-stakeholderism cannot prevent state intervention in the sense of deciding about the roles and competences of stakeholders, the right to participate in consultations or negotiations, and what elements of the agenda are discussed. At the same time, the Indigenous peoples of the Columbia Basin are not simply stakeholders or stewards with a right to natural resources management—they are also legally sovereign entities based on their inherent rights and nation-to-nation treaties. Thus, the modernization of the CRT is not a simple question of functionality and interests, or more efficient transboundary water management, but also that of ethics and respect.

Perhaps the case of the Columbia River Treaty renegotiation offers the insight that the role of Indigenous peoples in negotiating transboundary river issues may be approaching critical change. If the US and Canada accept that a new understanding of water diplomacy is necessary to decolonize the Columbia River Treaty, if they engage in pre-negotiation agenda setting and ask for the free, prior, and informed consent of their Indigenous peoples, if they negotiate in good faith to put tribes at and not just near the table, they can discover through practice how to foster reconciliation in a meaningful way. The renegotiation opens an opportunity to reframe the regional natural resource governance of the basin from a bilateral to a multilateral context (Cohen and Norman 2018) by expressly acknowledging the nations within and between the United States and Canada.

References

Anghie A (2006) The evolution of international law: colonial and postcolonial realities. Third World Q 27(5):739–753

Behn C, Bakker K (2019) Rendering technical, rendering sacred: the politics of hydroelectric development on British Columbia's Saaghii Naachii/Peace River. Glob Environ Polit 19(3):98–119

Bell DA (2015) Columbia River Treaty renewal and sovereign tribal authority under the Stevens Treaty "right-to-fish" clause. Public Land Resour Law Rev 36(8):269–294

Blakemore E (2022) US declares disaster for tribal salmon fisheries on the West Coast. Washington Post. https://www.washingtonpost.com/science/2022/09/10/tribal-fisheries-disaster. Accessed 22 Jan 2023

Bonneville Power Administration (2001) The Columbia River System inside story. https://www.bpa.gov/-/media/Aep/power/hydropower-data-studies/columbia_river_inside_story.pdf. Accessed 11 Jan 2023

British Columbia (2014) Columbia River Treaty Review Public Consultation Report. https://engage.gov.bc.ca/app/uploads/sites/6/2017/01/Columbia-River-Treaty-Review-Public-Consultation-Report-_March-2014-1.pdf. Accessed 22 Jan 2023

British Columbia (2019) Columbia River Treaty News, 16 Sept 2019. https://engage.gov.bc.ca/columbiarivertreaty. Accessed 22 Jan 2023

British Columbia (2022) Canada—U.S. negotiations: where we are and what's to come. Zoom webinar recording, 16 May 2022. https://engage.gov.bc.ca/columbiarivertreaty/2022/05/16/canada-u-s-negotiations-where-we-are-and-whats-to-come. Accessed 22 Jan 2023

British Columbia (2023a) Columbia River Treaty: the treaty. https://engage.gov.bc.ca/columbiarivertreaty/the-treaty. Accessed 22 Jan 2023

British Columbia (2023b) Minister's statement on 15th round of Columbia River Treaty negotiations. https://news.gov.bc.ca/releases/2023EMLI0005-000105. Accessed 22 Jan 2023

Brunnel DC (2017) Lessons learned from demise of Northwest aluminum industry. https://www.tricitiesbusinessnews.com/2017/04/aluminum-industry. Accessed 22 Jan 2023

CCRIFC (2016) History: Canadian Columbia River inter-tribal fisheries commission. http://ccrifc.org/history. Accessed 22 Jan 2023

Chunhabunyatip P et al (2018) Influence of indigenous spiritual beliefs on natural resource management and ecological conservation in Thailand. Sustainability 10(8):2842

Cohen A, Norman ES (2018) Renegotiating the Columbia River Treaty: transboundary governance and indigenous rights. Glob Environ Polit 18(4):4–24

Collins D (2019) Being heard is half the battle: tribes speak up for ecosystem needs during Columbia River Treaty negotiations. Columbia Insight. https://columbiainsight.org/being-heard-is-half-the-battle. Accessed 11 Jan 2023

Columbia Basin Tribes Coalition (2010) Common views on the future of the Columbia River Treaty. https://critfc.org/wp-content/uploads/2015/05/Common-Views-statement.pdf?x78172. Accessed 22 Jan 2023

Columbia Basin Trust (2018) Columbia River Basin: dams and hydroelectricity. https://thebasin.ourtrust.org/wp-content/uploads/downloads/2018-07_Trust_Dams-and-Hydroelectricity_Web.pdf. Accessed 11 Jan 2023

Columbia River Treaty Negotiations Framework Agreement, 15 June 2019. https://secwepemcstrong.com/wp-content/uploads/2021/02/CRT-Negotiation-Framework-Agreement_revised-1.pdf. Accessed 22 Jan 2023

Columbia River Treaty Review—B.C. Decision, Nov 2011. https://engage.gov.bc.ca/app/uploads/sites/6/2012/03/BC_Decision_on_Columbia_River_Treaty.pdf. Accessed 22 Jan 2023

Columbia Valley Pioneer (2019) Columbia River Treaty negotiators invite First Nations observers to speak. https://www.columbiavalleypioneer.com/columbia-river-treaty-negotiators-invite-first-nations-observers-to-speak. Accessed 22 Jan 2023

Conroy K (2022) A busy year for Columbia River Treaty negotiations. https://engage.gov.bc.ca/columbiarivertreaty/2022/12/14/minister-op-ed-a-busy-year-for-columbia-river-treaty-negotiations. Accessed 22 Jan 2023
Cotsirilos T (2022) Epic floods in Pacific Northwest revive long-running dispute over how to manage a river. Food & Environment Reporting Network. https://thefern.org/2022/01/epic-floods-in-pacific-northwest-revive-long-running-dispute-over-how-to-manage-a-river. Accessed 22 Jan 2023
Crawley T (2018) First Nations excluded from Columbia River Treaty talks. https://www.saobserver.net/news/first-nations-excluded-from-columbia-river-treaty-talks. Accessed 22 Jan 2023
CRITFC (2013) About spirit of the salmon: creation story. https://plan.critfc.org/2013/spirit-of-the-salmon-plan/about-spirit-of-the-salmon/creation-story. Accessed 11 Jan 2023
CRITFC (2021a) Fisheries timeline: chronology of tribal fishing and fishing rights on the Columbia River. https://critfc.org/about-us/fisheries-timeline. Accessed 11 Jan 2023
CRITFC (2021b) We are all Salmon People. https://critfc.org/salmon-culture/we-are-all-salmon-people. Accessed 11 Jan 2023
Deloria V Jr, Wilkins DE (1999) Tribes, treaties, and constitutional tribulations. University of Texas Press, Austin
Department of Justice Canada (2021) Principles respecting the Government of Canada's relationship with Indigenous peoples. https://www.justice.gc.ca/eng/csj-sjc/principles-principes.html. Accessed 11 Jan 2023
First Nations Development Institute, Echo Hawk Consulting (2023) Reclaiming native truth: a project to dispel America's myths and misconceptions. https://rnt.firstnations.org/?_ga=2.179407204.1107999603.1676473684-1897899823.1676473683. Accessed 22 Jan 2023
Fontaine T (2016) Canada officially adopts UN declaration on rights of Indigenous Peoples. CBC News. https://www.cbc.ca/news/indigenous/canada-adopting-implementing-un-rights-declaration-1.3575272. Accessed 22 Jan 2023
Garneau M (2022) Barriers to economic development in Indigenous communities: report of the Standing Committee on Indigenous and Northern Affairs. House of Commons, Canada. https://www.ourcommons.ca/Content/Committee/441/INAN/Reports/RP11714230/inanrp02/inanrp02-e.pdf. Accessed 22 Jan 2023
Global Affairs Canada (2019) Federal government announces Columbia River Basin Indigenous Nations to participate as observers in Columbia River Treaty negotiations. https://www.canada.ca/en/global-affairs/news/2019/04/federal-government-announces-columbia-river-basin-indigenous-nations-to-participate-as-observers-in-columbia-river-treaty-negotiations.html. Accessed 22 Jan 2023
Government of Canada (2022) Canada-US Columbia River treaty. https://www.canada.ca/en/environment-climate-change/corporate/international-affairs/partnerships-countries-regions/north-america/canada-united-states-columbia-river.html. Accessed 11 Jan 2023
Harrison J (2021) With few exceptions, Columbia River salmon and steelhead returns continue downward trend. Northwest Power and Conservation Council. https://www.nwcouncil.org/news/2021/03/12/few-exceptions-columbia-river-salmon-and-steelhead-returns-continue-downward-trend. Accessed 11 Jan 2023
Hemmati M (2000) Multistakeholder processes for governance and sustainability: beyond deadlock and conflict. Earthscan, London.
Hocking B (2006) Multistakeholder diplomacy: forms, functions, and frustrations. In: Kurbalija J, Katrandijev V (eds) Multistakeholder diplomacy—challenges and opportunities. DiploFoundation, Msida-Geneve, pp 13–32
Idaho Conservation League (2021) The Northwest's Future: solutions together. https://www.idahoconservation.org/our-work/salmon-and-steelhead/solutionstogether. Accessed 22 Jan 2023
International Columbia River Engineering Board (1959) Water resources of the Columbia River Basin: report to the International Joint Commission. International Columbia River Engineering Board, Ottawa
International Joint Committee (2022) International Columbia River Board of Control: Mandate. https://www.ijc.org/en/crbc/who/mandate. Accessed 22 Jan 2023

Jansen LJM, Kalas PP (2020) Improving governance of tenure in policy and practice: a conceptual basis to analyze multi-stakeholder partnerships for multi-stakeholder transformative governance illustrated with an example from South Africa. Sustainability 12(23):9901

Katrandijev V (2006) Reflections on multistakeholder diplomacy. In: Kurbalija J, Katrandijev V (eds) Multistakeholder diplomacy—challenges and opportunities. DiploFoundation, Msida-Geneve, pp 123–140

Larson AM et al (2022) A place at the table is not enough: accountability for indigenous peoples and local communities in multi-stakeholder platforms. World Dev 155:105907

Letter of Agreement on the Commitment to Collaborate on an Initiative relating to the objective of the re-introduction of Anadromous Salmon into the Canadian Portion of the Columbia River Basin, 29 July 2019. https://engage.gov.bc.ca/app/uploads/sites/6/2019/12/SALMON-LETTER-OF-AGREEMENT.pdf. Accessed 22 Jan 2023

Letter of the Columbia Basin Tribes Coalition to Brigadier General John S. Kem and Mr. Elliot E. Mainzer, 14 Feb 2014. https://naiads.files.wordpress.com/2014/02/fish-passage-white-paper-2-14-14.pdf. Accessed 22 Jan 2023

Lichatowich JR et al (2017) Wild Pacific salmon: a threatened legacy. Bemis Printing, St. Helens

Mapendere J (2008) Track one and a half diplomacy and the complementarity of tracks. Cult Peace Online J 2(1):66–81

Mapes LV (2021) GOP congressman pitches $34 billion plan to breach Lower Snake River dams in new vision for Northwest. Seattle Times. https://www.seattletimes.com/seattle-news/environment/gop-congressman-pitches-34-billion-plan-to-breach-lower-snake-river-dams-in-new-vision-for-northwest. Accessed 22 Jan 2023

Martens J (2007) Multistakeholder partnerships—future models of multilateralism? FES Dialogue on Globalization. https://www.globalpolicy.org/sites/default/files/martens_multistakeholder_partnerships_online_version.pdf. Accessed 22 Jan 2023

Mason MD (1983) Canadian and United States approaches to Indian sovereignty. Osgoode Hall Law J 21(3):422–474

Mclash RA (2018) Island voices: First Nations excluded from Columbia River Treaty negotiations. https://www.timescolonist.com/opinion/island-voices-first-nations-excluded-from-columbia-river-treaty-negotiations-4664455. Accessed 22 Jan 2023

NOAA Fisheries (2022) Rebuilding Interior Columbia Basin Salmon and Steelhead. National Oceanographic and Atmospheric Administration National Marine Fisheries Service. https://media.fisheries.noaa.gov/2022-09/rebuilding-interior-columbia-basin-salmon-steelhead.pdf. Accessed 11 Jan 2023

Northwest Power and Conservation Council (2022a) Dams: history and purpose. https://www.nwcouncil.org/reports/columbia-river-history/damshistory. Accessed 11 Jan 2023

Northwest Power and Conservation Council (2022b) Fish passage at dams. https://www.nwcouncil.org/reports/columbia-river-history/fishpassage. Accessed 22 Jan 2023

Pynn L (2021) Renegotiation of Columbia River Treaty offers hope for salmon, tribes. https://www.sixmountains.ca/article/d7dc2b07-8ba8-4a0c-8d7c-df64e64d1cca. Accessed 22 Jan 2023

Quaempts EJ et al (2018) Aligning environmental management with ecosystem resilience: a First Foods example from the Confederated Tribes of the Umatilla Indian Reservation, Oregon, USA. Ecol Soc 23(2):29–49

Rahr G (2016) Why protect salmon? Wild Salmon Center. https://www.waterboards.ca.gov/waterrights/water_issues/programs/bay_delta/california_waterfix/exhibits/docs/PCFFA&IGFR/part2/pcffa_180.pdf. Accessed 22 Jan 2023

Ramey D (2021) Tribal Nations pass resolution to back Idaho Rep. Simpson's plan to breach 4 Snake River dams. https://www.ktvb.com/article/news/politics/tribal-nations-pass-resolution-back-rep-simpson-plan-breach-snake-river-dams/277-4534851d-3551-49cf-82b3-d7fff9bfc1f4. Accessed 22 Jan 2023

Ratner B et al (2018) Investing in multi-stakeholder dialogue to address natural resource competition and conflict. Dev Pract 28(6):799–812

Reimer J (2021) Termination: a solution to Canadian entitlement valuation disputes. Oregon Rev Int Law 22:223–248

Rodriguez D, Sarmiento Barletti JP (2021) Can multi-stakeholder forums mediate indigenous rights and development priorities? Insights from the Peruvian Amazon. Int for Rev 23(1):103–115

Samala S (2021) The use of indigenous traditional knowledge in climate change strategies. Wilson Center. https://www.wilsoncenter.org/event/use-indigenous-traditional-knowledge-climate-change-strategies. Accessed 11 Jan 2023

Sarmiento Barletti JP et al (2022) Understanding difference to build bridges among stakeholders: perceptions of participation in four multi-stakeholder forums in the Peruvian Amazon. J Dev Stud 58(1):19–37

Save Our Wild Salmon (2023) Modernizing the Columbia River Treaty. https://www.wildsalmon.org/projects/modernizing-the-columbia-river-treaty/modernizing-the-columbia-river-treaty.html. Accessed 22 Jan 2023

Schafer T (2022) Return of salmon part of Columbia River Treaty negotiations. https://www.timeschronicle.ca/return-of-salmon-part-of-columbia-river-treaty-negotiations. Accessed 11 Jan 2023

Scruggs G (2019) Roll on, Columbia? U.S. tribes demand seat at river treaty negotiating table. Reuters. https://www.reuters.com/article/us-americas-river-salmon/roll-on-columbia-us-tribes-demand-seat-at-river-treaty-negotiating-table-idUSKCN1RT1FE. Accessed 22 Jan 2023

Sewell WRD (1964) The Columbia River Treaty: a landmark in international river development. Yearbook of the Association of Pacific Coast Geographers 26:15–22

Stern CV (2022) Columbia River Treaty Review. Congressional Research Service Report R43287. https://sgp.fas.org/crs/misc/R43287.pdf. Accessed 22 Jan 2023

Stuttaford A (2022) Damned if You Do… National Review. https://www.nationalreview.com/corner/damned-if-you-do. Accessed 22 Jan 2023

Syilx Okanagan Nation Alliance (2022) Columbia River Treaty Information and Updates. https://www.syilx.org/wp/wp-content/uploads/2022/08/CRT-Communique-2.pdf. Accessed 22 Jan 2023

The Columbia Basin Tribes & First Nations (2015) Fish passage and reintroduction into the U.S. & Canadian Upper Columbia Basin. https://ucut.org/wp-content/uploads/2016/09/Fish_Passage_and_Reintroduction_into_the_US_And_Canadian_Upper_Columbia_River4-1.pdf. Accessed 22 Jan 2023

The Columbia River Salmon Reintroduction Initiative (CRSRI) (2023) Five partners. https://columbiariversalmon.ca/about/five-partners. Accessed 22 Jan 2023

Thompson A et al (1996) Loss of the Columbia River First Nations fishery: review of the potential for legal action against the federal government. http://ccrifc.org/cms/wp-content/uploads/2015/10/Loss-of-the-Columbia-River-First-Nations-Fishery-Review-of-the-Potential-for-Legal-Action-Against-the-Federal-Government.pdf. Accessed 11 Jan 2023

Truth and Reconciliation Commission of Canada (2015) Calls to Action. https://www2.gov.bc.ca/assets/gov/british-columbians-our-governments/indigenous-people/aboriginal-peoples-documents/calls_to_action_english2.pdf. Accessed 22 Jan 2023

U.S. Bureau of Reclamation (2021) How has Grand Coulee Dam impacted salmon? https://www.usbr.gov/pn/grandcoulee/about/faq.html#salmon. Accessed 11 Jan 2023

U.S. Department of Justice (2022) Native Americans. https://www.justice.gov/otj/about-native-americans. Accessed 11 Jan 2023

U.S. Department of State (2011) Announcement of U.S. support for the United Nations Declaration on the Rights of Indigenous Peoples. https://2009-2017.state.gov/s/srgia/154553.htm. Accessed 22 Jan 2023

U.S. Department of State (2018) On the opening of negotiations to modernize the Columbia River Treaty regime https://2017-2021.state.gov/on-the-opening-of-negotiations-to-modernize-the-columbia-river-treaty-regime/index.html. Accessed 22 Jan 2023

U.S. Entity Regional Recommendation for the Future of the Columbia River Treaty after 2024, 13 Dec 2013. https://www.bpa.gov/-/media/Aep/projects/columbia-river-treaty/crt-regional-recommendation-12-13-13.pdf. Accessed 22 Jan 2023

UNGA International Law Commission (2019) First report on the general principles of law. A/CN.4/732. https://documents-dds-ny.un.org/doc/UNDOC/GEN/N19/100/93/PDF/N1910093.pdf?OpenElement. Accessed 11 Jan 2023
United States Entity (2012) Columbia River Treaty 2014/2024 review: recent study results. Accessed through Wayback Machine, 22 Jan 2023
United States Entity (2013) Columbia River Treaty 2014/2024 review: Canadian entitlement. Accessed through Wayback Machine, 22 Jan 2023
US Army Corps of Engineers, Bonneville Power Administration (2009) Columbia River Treaty: history and 2014/2024 review. https://usace.contentdm.oclc.org/digital/collection/p16021coll7/id/8409. Accessed 22 Jan 2023
Walter E et al (2000) Fishing around the law: the Pacific salmon management system as a "structural infringement" of aboriginal rights. McGill Law J 45:263–314
Weaver T (1997) Litigation and negotiation: the history of salmon in the Columbia River Basin. Ecol Law Quart 24(4):677–687
Ween GB, Colombi BJ (2013) Two rivers: the politics of wild salmon, indigenous rights and natural resource management. Sustainability 5:478–495
Yachnin J (2022) Republicans reject Lower Snake River dam-breaching plan. E&E Daily. https://www.eenews.net/articles/republicans-reject-lower-snake-river-dam-breaching-plan. Accessed 22 Jan 2023

Chapter 10
The Role of Data Sharing in Transboundary Waterways: The Case of the Helmand River Basin

Najibullah Loodin, Gabriel Eckstein, Vijay P. Singh, and Rosario Sanchez

Abstract While data and information exchanges theoretically play an effective role in the decision-making process of a shared watercourse, in practice, there are several challenges that prevent riparians from sharing data in an effective and cooperative manner. This chapter seeks to assess why the riparian nations of the Helmand River have failed to adopt an effective data exchange mechanism although both nations signed an internationally recognized bilateral water treaty in 1973. Applying a mixed study approach, the study draws on the theory of planned behavior (TPB) to interpret the main obstacles of data sharing between Afghanistan, the upstream state, and Iran, the downstream state, of the Helmand River Basin. Using both secondary literature data and stakeholder interviews, this research found a number of specific factors that impede the ability of the riparian nations of the Helmand River to share data and information, such as national security, political instability, lack of capacity, negative emotions, pride, and historical factors. Our research suggests that these underlying elements have been under-estimated in the TPB models of data sharing at the transboundary level. Thus, borrowing from Wehn's model (2003b), we have developed a modified version of TPB to unpack the challenges of data sharing between the riparian states of the Helmand River Basin.

N. Loodin (✉)
Water Management and Hydrological Science Program, College of Arts and Sciences, Texas A&M University, College Station, TX, USA
e-mail: loodin2234@tamu.edu

G. Eckstein
Texas A&M University School of Law, College Station, TX, USA
e-mail: gabrieleckstein@law.tamu.edu

V. P. Singh
Department of Biological and Agricultural Engineering and Zachry Department of Civil and Environmental Engineering, Texas A&M University, College Station, TX, USA
e-mail: Vijay.Singh@ag.tamu.edu

R. Sanchez
Texas Water Resources Institute, Texas A&M University, College Station, TX, USA
e-mail: Maria.SanchezFlores@ag.tamu.edu

K. Szálkai and M. Szalai (eds.), *Theorizing Transboundary Waters in International Relations*, Springer Water, https://doi.org/10.1007/978-3-031-43376-4_10

Keywords Helmand River · Hamoun · Water conflict · Chahnimehs · Data sharing

10.1 Introduction

Managing an international watercourse is a complicated process, where various actors with different interests can influence the flow of water within the basin (Botterweg and Rodda 1999; Toset et al. 2000; Wolf et al. 2003). In fact, the management of a transboundary river basin can be challenging, as states can adopt unilateral resource-capturing strategies for exploiting the shared water, both in quantity and in quality. Such unilateral resource-capturing policies can lead to water conflicts between the riparian states (Sadoff and Grey 2002; Furlong et al. 2006; Gleditsch et al. 2006). However, proper data and information exchange may prevent water conflicts between the upstream and downstream nations of a shared river basin (Barua et al. 2018).

Using a case study exploring the Helmand River Basin, shared between Afghanistan and Iran, this chapter borrows from the Theory of Planned Behavior (TPB), according to which planned behavior is someone's intended assessment (either positive or negative) of the self-performance of a peculiar behavior (Kerr et al. 2005). Scholars such as Pooreh and Nodeh (2015) posit that planned behavior is "an individual's perception about the particular behavior, which is influenced by the judgment of significant others (p. 840)," such as the policy makers of a shared river basin. TPB was developed by Ajzen in 1985, and it is widely applied to data sharing, for example of hydrological data, at transboundary level, including the Mekong River Basin, which is shared between China, Thailand, Vietnam, Laos, Myanmar, and Cambodia.

However, to further unpack the challenges of data sharing in the basin, we have modified the TPB model developed by Wehn (2003a, 2003b). We felt that Wehn's TPB model under-estimated some important factors that highly influence the data sharing mechanism of riparians on a shared watercourse. Applying a mixed approach, we relied on both secondary data and semi-structured interviews conducted with officials, scholars, and authorities from both Afghanistan and Iran, as well as two international scholars who were previously engaged in the basin. In accordance with good research ethics practice, the study uses pseudonyms for all interviewees.

The chapter first presents the theoretical framework of the study, followed by the introduction of the historical background to the case study. The results of the analysis are then explored, followed by a discussion of their implications. The study concludes that despite the benefits of data sharing in an international shared water-course, there are some underlying factors that have prevented both Helmand River riparians from cooperating over data and information. These factors, which were under-represented by scholars such as Wehn (2000, 2001, 2003a, 2003b), Thu and Wehn (2016), Gharesifard (2015), and Gharesifard and Wehn (2016), are emotions, history of a shared river basin, pride, national identity, and securitization.

10.2 Theoretical Framework—The Role of Data Sharing in Transboundary Water Relations

10.2.1 The Importance of Data Sharing from a Theoretical and Empirical Perspective

The importance of data sharing in transboundary water management can be verified by both theoretical and empirical analysis. From a theoretical point of view, it can be argued that sharing scientific information is the foundation for the successful negotiation of agreements between riparian states (Gerlak and Grant 2009). Thu and Wehn (2016) believe that sharing information on a transboundary watershed among riparian states can help mitigate the negative impacts of droughts and flooding. Moreover, data sharing provides the basis for assessing the impacts of socio-economic activities on water resources (Timmerman and Langaas 2004, 2005), enhancing the functionality of ecosystems, and increasing adaptive measures in response to climate change (Wilby and Dessai 2010; Singh 2017; Singh et al. 2014).

Neoliberalism, a school of International Relations, is not only an economic ideology, it is also widely used in political ecology. It describes the relationship between power and knowledge sharing that can encompass data sharing at the international basin level (Ong 2006). The neoliberalism school of thought centers on the discourse of data openness and free data availability. For example, through the availability of public data at the inter-basin level, riparian nations, especially the local communities in a shared river basin, will witness economic growth and stability in the basin (Holmwood 2017). Additionally, data sharing at the international basin level leads to transparency, accountability, and fairness in the utilization of natural resources, which in our context refers to the water flow of a shared waterway (Slavnic 2017).

Thus, we first apply the neo-liberal theory to assess why the riparian nations of the Helmand River Basin lack cooperation and how this challenges the sustainable utilization of the river. While some riparian countries of a transboundary watercourse prefer cooperation over non-cooperation, other riparian countries tend to utilize the resources in an unfair and unequitable manner, triggering water conflict between the neighboring countries (Keohane and Nye 1973). We then update the neo-liberal model with components of constructivist theory, through which we argue that for the enhancement of water cooperation between the riparian nations of the Helmand River Basin, both upstream and downstream states must commit to data sharing. This is because the southern part of Afghanistan and the Sistan-Baluchistan region of Iran rely on the sustainable utilization of the Helmand River, which could be achieved through a reliable and thorough data sharing mechanism between the two countries (Thies 2004).

Additionally, data sharing is considered a means for building trust between riparian states, which eventually leads to strong water cooperation among the riparian states (Kliot and Shmueli 2001; Kliot et al. 2001; Eckstein 2005). Although many riparian states sign international water agreements on specific international basins that enable them to share data and information, the mechanisms for accessing reliable data on a shared transboundary basin are challenging (Gerlak et al. 2011), especially when climatic changes and over-utilization in both upstream and downstream nations negatively impact river flows in terms of both quantity and quality (Armitage et al. 2015).

Data sharing is crucial for the sustainable and transparent management of a transboundary watercourse. The thirteen interviewees who were interviewed for this research project, for example, overwhelmingly supported the value of data sharing. Eight out of the thirteen stakeholders interviewed for the project emphasized the need for regular data sharing. Out of eight, specifically, five interviewees advocated annual data sharing while the rest emphasized semi-annual data sharing. Moreover, many of them highlighted the importance of event-triggered data sharing, especially during flood and drought periods. In addition, the interviewees highlighted the value of data sharing in many different fields, including seven main topics they consider especially important for transboundary basin data sharing. These priorities include information on water quality, water pollutants, groundwater abstraction, and future planned measures in the basin, as well as metadata on methodologies and approaches, meteorological data, and hydrological data.

Data sharing can take place in different forms and through different mechanisms. Beside the traditional intergovernmental level, many authors (Chenoweth and Feitelson 2001; Schmeier et al. 2016; Porta and Wolf 2021) recognize joint research centers and river basin organizations as optimal tools in transboundary river management, as well as their contribution to water data and information exchange between riparian states. Moreover, they also have the potential to improve mutual trust between the riparian states of a shared watercourse (Burton and Molden 2005; Susskind and Islam 2012; Sarfaraz et al. 2022). These theoretical expectations were echoed by the interviewees, who also highlighted the importance of establishing an independent joint research center where scholars, policy makers, and government staff (as well as independent researchers) could jointly engage in researching the shared river basin.

10.2.2 Explaining the Shortcomings in Data Sharing—The Theory of Planned Behavior (TPB) Model

While both researchers and practitioners agree that data and information exchange may lead to mutual benefits, in practice, many obstacles can arise. One of the main challenges is the cost of developing the data and information resources. A lack of mutually credible data and information can undermine any effort of data sharing. As this case study will show, facing political or economic fragility, riparian states might not be able to either develop reliable data or participate in joint data development. Moreover, as Beniston et al. (2012) argue, in some cases data politization and manipulation, the high cost of data collection, license issuance for data collection, and national regulations might also add to the problem of having reliable information, and eventually, data sharing.

Researchers also identify several other reasons why data sharing is not achieved. Chenoweth and Feitelson (2001) suggest that the main obstacles for data and information exchange are differences in legal systems, mistrust, language, culture, technical capacity, economic growth and development, and an increasing demand for water use among the riparian states of a shared river basin. Past history of conflict in a shared river basin can also adversely affect the transboundary dialogue between the riparians of that basin. In fact, the conflict background of a basin may trigger riparian states to favor tensions over shared interests (Brady 2017).

In addition, the scarcity of water in a transboundary basin can inevitably push riparians to consider water as a national security interest, which can also create challenges for data exchange (Dormido 2019; Prasai 2013; Faizee 2022). Shroder (2016) argues that considering water a component of national security will impede the equitable and reasonable utilization of water among riparian states (p. 342). Thus, tying water to national security can also hinder data exchange procedures in water-stressed countries (Plengsaeng et al. 2014).

While the above-mentioned factors can serve as impediments to data sharing between riparian states, we argue that there are other interrelated and relevant factors that can influence data exchange mechanisms at a transboundary level. These include emotions, history and pride, national identity, and securitization (Hutchison 2016; Sultana 2011; Seide 2010, 2016, 2018).

To address the challenges of data sharing in the Helmand River Basin in a comprehensive manner, and to incorporate these additional factors into the analysis, we began with the Theory of Planned Behavior first developed by Ajzen (1985), who contended that TPB helps understand how people at the decision-making level (e.g. officials) behave. TPB has been widely applied in data sharing at the transboundary level. For instance, Gharesifard (2015) and Gharesifard and Wehn (2016) apply it in Italy, the UK, and the Netherlands, while Thu and Wehn (2016) and Plengsaeng et al. (2014) apply it to the Mekong River Basin. We in turn applied this theory to assess the behavior of Iran and Afghanistan in the context of data sharing over the Helmand River Basin. The model assumes that behavior is planned, and so it is easy to anticipate the intentional behavior of riparian states in sharing or not sharing water data in a transboundary basin (Wehn 2000, 2001, 2003a, 2003b).

The components of TPB as stated by Ajzen (1985) are as follows:

1. **Attitude** shows the friendly (pleasing) and unfriendly evaluation of the behavior of interest (Cherry 2022).
2. **Subjective Norms** highlight how people approve or disapprove of each other's behavior (Gawronski 2007), in our case, whether the riparian states of Helmand River approve or disapprove of each other's contentious or cooperative manner.
3. **Perceived behavior control** shows people's varying perception of how challenging or easy the performance of the behavior of interest is (Pooreh and Nodeh 2015).

4. **Behavior intention** highlights the factors that shape or co-shape a behavior. It is also argued that the more powerful the intention, the more certainly the behavior will be performed (Sheeran and Webb 2016).
5. **Behavior** is "an attempt on the part of an individual to bring about some state of affairs – either to effect a change from one state of affairs to another, or to maintain a currently existing one" (Bergner 2011, p. 148).

However, one limitation of TPB, according to Wayne LaMorte (2022), is that the theory under-values the role of interrelated factors, especially the role of emotion and how emotion influences the decision-making processes of the policy makers of the riparian states in an internationally shared river basin. Thus, we argue that to further assess the challenges of water cooperation in the context of the Helmand River Basin, the TPB model should be revised by including factors such as emotion, history, pride, and national identity in Wehn's TPB model (2003a, 2003b) because these factors shape the decision-making processes of the riparian countries related to data sharing on the transboundary Helmand River Basin (see Fig. 10.1).

We argue that the model explains the motivations behind data sharing in specific cases, or the lack of it. To engage in the practice of data sharing, states have to be committed to do so. This commitment comprises three components:

a. **Attitude**: Attitude demonstrates the viewpoint or position of each authority (i.e. both the upstream and the downstream riparian countries) about the positive and negative impacts of data sharing.
b. **National Interest(s)**: This notion is deeply influenced by culture, tradition, social values, ideology, pride, national identity, and security (Liu 2013). In the context of the Helmand River Basin, water is considered to be a national security priority for both Iran and Afghanistan, which is embedded in emotions, history, and pride (Voice of America Persian [VOA] 2022; Shroder 2016). Iceland and Otto (2017) contend that when thinking about national security issues, rivers, lakes,

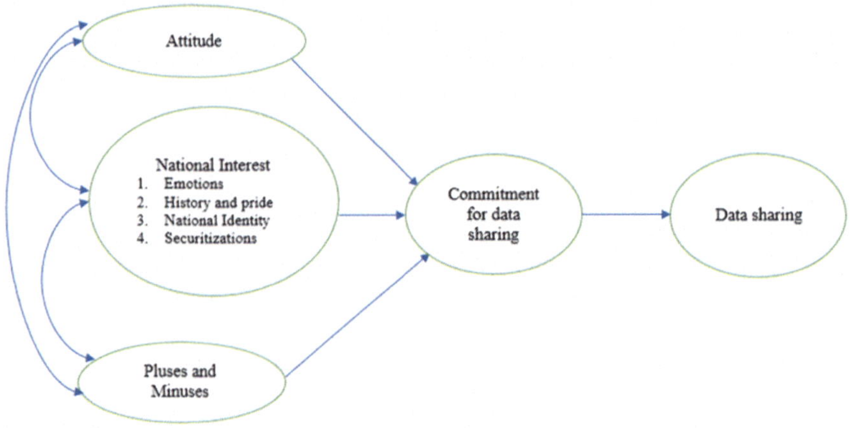

Fig. 10.1 Proposed conceptual model for data sharing. Created by the authors

and glaciers are the common issues that leap into our mind. Water is typically considered a critical national security concern in water-stressed countries that suffer from climate change and drought (Tadros 1996).

b.1. **Emotion**: Emotion is a mix of socio-biological reactions that influence the activities of human beings. As a biological reaction, emotion prepares the body to act in any situation (Coelho et al. 2016). Myers and Smith (2012) describe emotions as subjective feelings that trigger someone to feel in a specific manner, e.g. feeling anger or joy. As a social phenomenon, emotion is established by postures, gestures, and facial expressions that involve feelings, body movement, and non-verbal communications (expressions) to prepare human beings to react to their surroundings (Coelho et al. 2016). In international relations, emotion is an under-represented field of study. Emotion has various definitions, depending on whom we ask (Gendron 2010). As a biological state, emotion occurs due to our feelings, behaviors, and thoughts attached to something. While Davis (n.d.) categorizes emotion into pleasure and displeasure (satisfaction and dissatisfaction) subsets, Hutchings et al. (2022) describe emotion as positive and negative. For instance, positive emotion leads to hope and cheerfulness, which benefits health and the socio-economic condition of a community; in contrast, a negative emotion such as sadness, anger, or anxiety has a negative impact on health and the well-being of communities (see Frederickson and Joiner 2002). In international relations discourses, political emotion has received less attention, although during a crisis of natural resources (famine, drought, or flooding), emotion plays a significant role in the formation of politics in the community (Hutchison 2016).

In water-stressed countries such as Iran (Foltz 2002) and Afghanistan (Dormido 2019), access to shared water is contested. For example, in the Helmand River Basin, one of the most arid basins in the region with less than 100 mm annual precipitation, access to shared water is highly challenging. The right to access this limited quantity of water is embedded within emotions of anger, grief, and sometimes even threats from both upstream and downstream states. Since the two countries became independent states, the number of water conflicts between the riparians has increased significantly. This became especially evident when the Afghan government decided to resume the construction of Kamal Khan Dam in Chahar-Burjak District, Nimrouz Province, in the lower Helmand River Basin in Afghanistan (Kakar and Sayedi 2022; Saif 2021; Mehdi 2021; Ramachandran 2022; Askari and Bashardost 2021). Once construction started, Iranian officials expressed their deep concerns over the flow of the Helmand River below the dam. For instance, former President of Iran Hasan Rouhani stated that "we cannot remain indifferent to the [water dams] issue which is apparently damaging our environment." He further said that "Construction of several dams in Afghanistan, such as Kajaki, Kamal Khan, Salma and others in the north and south of Afghanistan, affect our Khorasan and Sistan-Baluchistan provinces" (Zahid 2017). It can be argued that this statement is embedded within the emotions of anger and grief, and even threat.

b.2. **History and pride**: Historians and philosophers assess pride both positively and negatively. For example, while Hume defines pride as an "agreeable impression, which arises in the mind, when the view either of our virtue, beauty, riches or power makes us satisfied with ourselves" (Hume 1896, pp. 277, 297), Aristotle argues that pride (satisfaction that someone gains from something or somewhere, such as a country or region, e.g. the Sistan region) can be achieved through doing or achieving great things (Smith 1998). He also asserts that "the man is thought to be proud who thinks himself worthy of great things, being worthy of them." However, Aristotle believes that to be proud of something, the man must be good in the highest degree (Smith 1998, p. 73). Historically speaking, the region encompassing the cities of Ghazni, Zabul, Kandahar, Bust, Zaranj, and Farah (all in Afghanistan), alongside Zabol in Iran, were collectively called Sistan. In terms of water management, Sistan was one of the most prominent basins in ancient and medieval times. Using community water management approaches, people in the region used to utilize the natural flow of water in a sustainable and reasonable way (Spandyar 2020; Bidabadi and Afshari 2020). In more recent times, the increase in population, coupled with excessive use of water for agricultural development, has largely contributed to a water conflict between the two riparian countries, especially since the political border between the two countries was drawn along the river in the 1870s. While Afghanistan did construct the Kamal Khan Dam in Nimrouz Province in 2021, the Afghan government has never initiated or operationalized any other large-scale dams due to its poor economic circumstances (Shroder 2016; Zahid 2017). Nevertheless, Iran has consistently expressed its extreme concern over the flow of the Helmand River. Like Egypt, as the downstream state on the Nile River (Seide 2010, 2018), Iran has repeatedly claimed historic rights over the flow of the Helmand River for agricultural, instream flow, and domestic uses (Nagheeby and Rieu-Clarke 2020).

The Sistan and Baluchistan sub-basin of the Helmand River continues to remind Iran of its ancestral great Persian (Achaemenid) Empire, and it is a source of great pride for Iranians. This region, which has been inhabited for more than 5,000 years, has been drained by the continuous flow of the Helmand River for centuries (Bidabadi and Afshari 2020). One of the prominent archeological sites in Sistan-Baluchistan in Iran is Shahr-i Sokhta, known as the Burnt City, which was established along the Helmand River in 3100 B.C. This city was home to people who used to make a living through community-led agriculture, most probably fruit cultivation (UNEP 2006). Iranians remember their great civilization in the basin and constantly assert historical and ancestral rights over water use in the basin. That is why the history of Sistan-Baluchistan is interwoven with Iran's national pride and interests. Policymakers in Iran argue that Goldsmid's arbitration on the demarcation of the political border of Afghanistan and Iran in the 1870s was a big mistake because it was detrimental to Iran and its historical claims to the region (Dabiri 2011). Despite the excessive use of water, climate change, and recurring droughts in both states (Mianabadi et al. 2020; Bhattacharyya et al. 2004; Lane 2020), Iran still demands its historic rights and believes it is entitled to the flow of the Helmand River as it flowed for centuries without any restriction. Similar to the Egyptian view on the Nile River (Sandstrom et al. 2018), ignoring the historic rights of Iran is a direct threat to the country's national interests. In contrast, the Afghan government, both the current Taliban-led regime and the previous government under President Ashraf Ghani, is committed to the implementation of the 1973 Afghan-Iranian Helmand River Water Treaty (Gul 2021; Mehdi 2021). Thus, the phenomenon of the pride and history that is embedded in the national interests of riparian countries is a challenge for water cooperation among co-riparian countries (Price et al. 2014).

b.3. **National Identity**: Smith defines national identity as a means for influencing individuals (e.g. the citizens of a region or the upstream or downstream users of a shared watercourse) and how individuals attach and express their feelings and emotions related to a country or region (Smith 1991). Calhoun (1997) argues that national identity assumes special priority over other collective identities in the construction of personal identity (p. 125). The notion of national identity is common in contemporary political geography. The incorporation of social theories into political geography has motivated geographers to explore various aspects of everyday life and how these aspects are influenced by political processes. These political processes affect people's lives in a way that awakens their sense of identity in a society (Storey 2009). Strictly speaking, national identity can be viewed based on a belief in common ethnicity, race, language, etc. (Tolia Kelly 2009), while loosely taken, other factors such as homes, villages, cities, natural resources, e.g. a river, can be considered a part of the national identity of a nation (Edensor 2009; Boelens 2015). For instance, the flow of the Nile River, which is shared between upstream Ethiopia and downstream Egypt (as well as other riparian states), is interwoven within the culture, language, religion, and identity of each country (Seide 2010). For thousands of years, people along the river, especially Egyptians, have utilized the waters of the Nile for irrigation, as well as cultural and spiritual affairs (Ahmed et al. 2020). Like with historic and present Egyptian civilizations, water has been used for spiritual and agricultural purposes in greater Persia (today's Iran, Afghanistan, Tajikistan, and Uzbekistan), where water is part of the national identity (Foltz 2002). In fact, Iranians perceive that the Sistan region, where the Helmand River flows (currently Nimruz province in Afghanistan and Sistan-Baluchistan in Iran), is an inseparable part of greater Persia (Dabiri 2011). Considering water as part of their national identity, they argue that at least two-thirds of the Helmand River flow should have been allocated to the residents of the Sistan-Baluchistan Region, as two-thirds of the Sistan region is situated in Iran (Dabiri 2011). In contrast, Shroder and Ahmadzai (2016) accuse Goldsmid of unfairly arbitrating the Sistan region between Afghanistan and Iran. Considering the Sistan region as their national identity, they argue that Goldsmid, the British General who was responsible for dividing up the Sistan Region between the two neighboring countries, arbitrated the Sistan region in favor of Iran as the British fought and lost the war with Afghans in the nineteenth century (p. 350).

c. **Pluses and Minuses**: Pluses and minuses refer to the ease and difficulty of data sharing at the transboundary level. Factors that contribute to or create challenges for data sharing include technical expertise, financial resources, type of data, method of sharing, data security, and opportunities (see Wehn 2003b). Once these barriers and challenges are overcome, riparian states are more likely to commit to data sharing.

10.3 Historical Background of the Tensions Regarding the Helmand/Hirmand River

The Helmand River Basin (known as the Hirmand in Iran) is the largest closed arid river basin in southern Afghanistan. This basin is drained by the Helmand River, which has supported various civilizations in the Sistan Valley, located in the lower part of the basin in southwestern Afghanistan, for more than 6,000 years (Dominguez et al. 1951; Spandyar 2020). The total area of the Helmand River Basin is approximately 400,000 square kilometers, covering almost 40% of the total area of Afghanistan. While 89% of the basin is located in Afghanistan, 9% of the basin is found in Iran, and 2% is located in Pakistan[1] (Mianabadi et al. 2020). The Helmand River originates in the Koh-I Baba Mountains in Central Afghanistan. The river begins its 1,300-km journey west of Kabul and flows southwest toward Iran (Najafi and Vatanfada 2013).

Before reaching the border, the river divides at a point known as the Helmand Fork. The northern part of the fork is called Chila-e Charkh River (known as the Parian in Iran) and flows northward, forming the political border between Afghanistan and Iran. The southern part of the fork flows into Iran, forming the Sistan River, which finally reaches Hamoun Helmand (Hamoun Hirmand in Iran), a lake and wetland area at the terminus of the Sistan River (Hearns 2015). According to some scholars, in a normal water year, the Helmand River used to flow into the hamouns (inland desert lakes and wetlands) that cover the border region in a counterclockwise direction (Shroder and Ahmadzai 2016). The river first flowed into Hamoun Puzak in Afghanistan, then to Hamoun Saberi (a transboundary hamoun located at the border of Afghanistan and Iran), and then into Hamoun Helmand (Whitney 2006; Shroder 2014). Moreover, when Hamoun Helmand overflowed, its waters ran into a small

[1] As only a small portion of the basin is situated in Pakistan, which is not part of the 1973 Helmand River Treaty, Pakistan is excluded from this analysis.

hamoun called Gowd-i-Zerah, located inside Afghanistan (McMahon 1906, pp. 218, 223) (see Fig. 10.2). However, all four of these hamouns have dried up ever since Iran built four Chahnimehs (artificial lakes) in 1981, just beyond the Helmand Fork in Iran (Goes et al. 2015, p. 18) to supply water to the Iranian cities of Zahedan and Zabol (Sharifikia 2013, p. 207). Today, a canal diverts the water from Helmand Fork to the four Chahnimehs, which has led to the desiccation of the hamouns in both Afghanistan and Iran (Van Beek et al. 2008).

Although there is no territorial dispute between the two riparian countries, Afghanistan and Iran disagree over the distribution of the waters of the Helmand River. Water tensions in the region date back to the 1870s, when British officer Sir Frederic Goldsmid drew the political border of the two countries along the Helmand River (Dominguez et al. 1951; Whitney 2006). The water conflict further escalated when the Afghan government decided to build the Kajaki Dam in the 1950s. To mitigate these tensions, in the 1950s and 1960s, the Afghan government proposed, as a

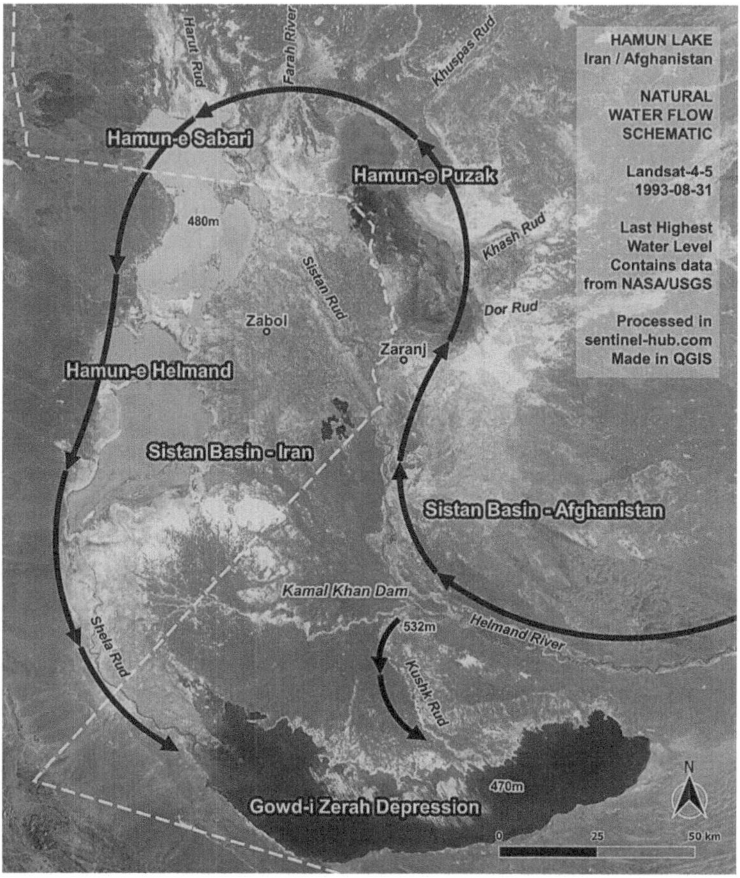

Fig. 10.2 Location of four hamouns in the lower Helmand River basin (Soar Earth 2022)

gesture of goodwill, that Afghanistan would deliver to Iran a supplementary 4 m^3/s of water in addition to the average of 22 m^3/s that the Helmand River Delta Commission recommended in 1951 (Faizee 2022; Loodin and Wolf 2022). This gesture, however, was conditioned on Iran allowing Afghanistan to expand its trade activities, both import and export, with other neighboring countries through the Chabahar and Bandar Abbas Ports in Iran (Abidi 1977).

On 21 March, 1973, the Prime Minister of Iran, Amir Abbas Hoveida, and the Prime Minister of Afghanistan, Musa Shafiq, signed the Afghan-Iranian Helmand River Water Treaty, in which Afghanistan accepted to allocate an average of 26 m^3/s of water from the Helmand River to Iran in a normal water year (Mianabadi et al. 2020).[2] In total, Iran's annual water share of the Helmand River is approximately 820.5 million cubic meters per year in a normal water year (Dabiri 2011; Thomas et al. 2016). The agreement also created a comprehensive system for managing transboundary water issues. According to Article III (A) of the treaty, water should be delivered to Iran in the Helmand River at three points:

1. At the place where the boundary line crosses Rude Sistan (Rude means river in Persian)
2. At pillar (marker) 51
3. At pillar 52 (see Hearns 2015; Dabiri 2011).

The treaty also stated that the commissioners of both riparian countries should jointly establish pillars 51 and 52. However, the above-mentioned domestic political transformation in the two states, as well as a lack of political will by authorities from both countries, has impeded the creation of the two pillars, leading to the intensification of disputes over the Helmand River (Ramachandran 2022; Climate Diplomacy 2016; Shroder and Ahmadzai 2016).

While the agreement was ratified both by Iran (1973) and Afghanistan (1977) (Abidi 1977), several circumstances have hindered its implementation. Soon after signing the treaty, Afghanistan began experiencing political instability, which has continued into the present, while Iran also underwent serious political transformation with the Islamic Revolution taking place in 1979–1980 (Milani 2018). Notably, the instability in Iran was coupled with its hydraulic mission to capture as much water as possible through the construction of dams, canals, and channels, an approach taken by some riparian nations (Wester et al. 2009). These factors have undermined the potential for cooperation between the two states.

Another tangible consequence of the domestic turmoil in the region occurred in Afghanistan from the mid-1970s, in that the country's water infrastructure, including the water management systems and structures on the Helmand River, drastically deteriorated. The hydrometric stations and other structures that were installed during the late 1950s and 1960s were destroyed during the internal war between the Mujahideen and the Soviet Union, and later between the Mujahideen and the Taliban (Sidky 2007;

[2] A normal water year, as defined in article 1 (c) of the Helmand River Treaty, is a year-long time period during which the flow of water at Dehrawud Station, located at Kajaki Dam, is at least 5,661 million cubic meters from 1 October to 30 September (Helmand River Treaty 1973).

Land Links 2018; World Bank 2018). Goes et al. (2015) argue that the lack of data between 1980–2004 is due to conflict and insecurity, which has especially challenged water resources assessment and management in the upper Helmand River Basin (also Shroder 2014, p.147).

The lack of discharge data from gauging stations is now responsible for an ongoing rift between Afghanistan and Iran over water rights and allocations under the 1973 treaty. Iran has consistently blamed the Afghan government for alleged shortfalls in the flow of the Helmand River and accused Afghanistan of not delivering Iran's rightful share of the water as provided for in the 1973 treaty (Amiri et al. 2013; Amini et al. 2021). In response, the Afghan government has claimed that Iran received more water than it was entitled to under the treaty over many years (Shroder 2016; Shirdeli 2014).

Using remote sensing data, Iranian scholars such as Mianabadi et al. (2020, 2021) argue that precipitation did not decline significantly in the basin between 1983 and 2016, contradicting Afghanistan's claims of low flows due to climate change and extreme drought especially between 1993–2001 (see Alami and Tayfur 2022; Mayar 2022) and during 2018, and 2021–2022 (Mayar 2022). As a result, Mianabadi et al. (2020) contend that Afghanistan has failed to deliver the amount of water that was agreed upon in the 1973 treaty. However, the authors acknowledge that their analysis lacks data on infiltration and evaporation, especially during the summer, when evaporation increases, as well as snowmelt data. Moreover, their analysis lacks reliable information, as they relied only on satellite data rather than actual flow measurements in the Helmand River (Mianabadi et al. 2020, p. 3459).

In contrast, the Water Resources Department of the Ministry of Energy and Water of Afghanistan (WRD-MEW 2022), the country's main agency for hydrological data collection and analysis, claims that on average, between 1973–1979 and from 2010–2020, Iran received more than double its water rights under the treaty based on measured discharge rate data from Khwabgah Station on the Helmand River (see Figs. 10.3 and 10.4).[3] Similarly, Sadat and Sayed (2020) estimate that "Iran has received 40% more water than its total agreed rights of 26 cubic meters per second per normal year since the signing of the treaty." Moreover, Shroder (2016) argues that Afghanistan has delivered more water to Iran, even in drought years, despite the country suffering from a lack of water infrastructure and because it simply does not use much of the water that it is entitled to (p. 469). Even with the construction of Kamal Khan Dam, farmers are still unable to irrigate their lands, as the necessary water infrastructures, such as diversion channels, have not been built yet (Kakar 2021).

Beside the domestic political and technical issues, climate change and severe drought have also significantly impacted the flow of the entire basin in both countries (Iqbal et al. 2018; Loodin and Warner 2022; Mayar 2022). Analyzing climatic conditions in the basin between 1970–2006, Roodari et al. (2021) argue that the Helmand River Basin suffered from drier weather due to a changing climate between

[3] Afghanistan also collected data between 2004–2010 at the Khwabgah and Chila-e Charkh Stations, however, according to the World Bank, that data is not reliable (World Bank 2018).

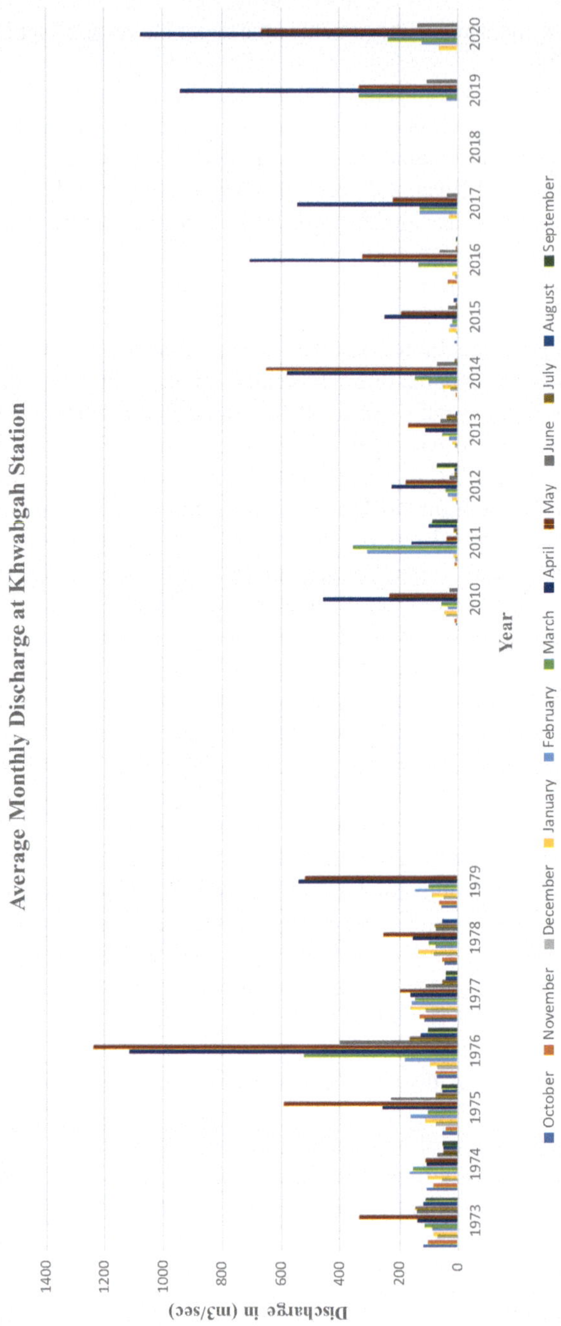

Fig. 10.3 Average monthly discharge at Khwabgah Station (WRD-MEW 2022)

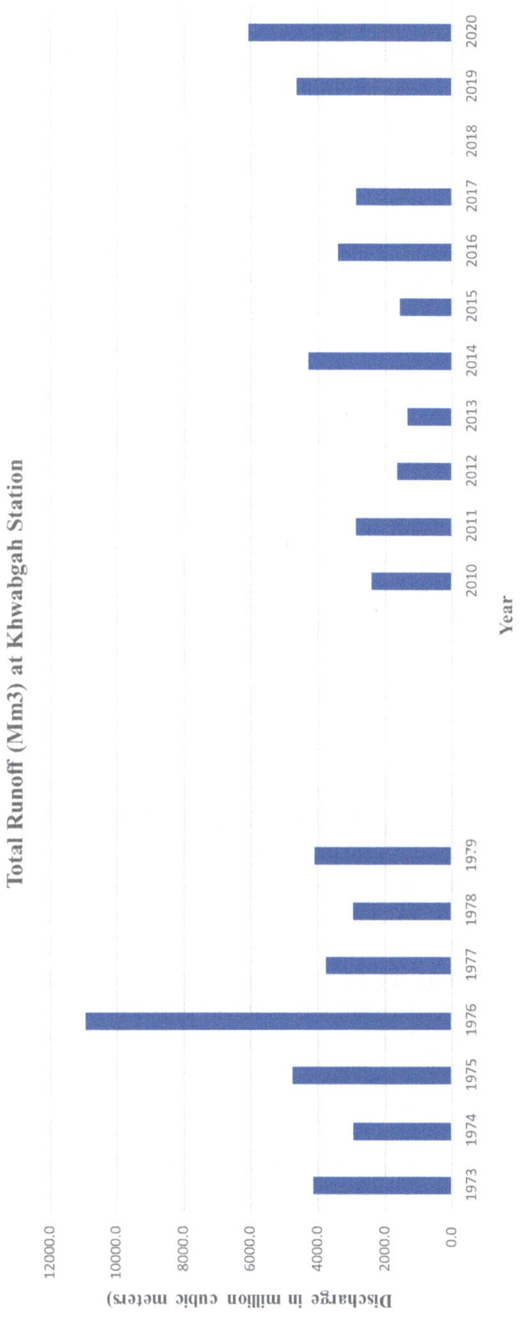

Fig. 10.4 Total runoff (Mm3) at Khwabgah Station (WRD-MEW 2022)

1970–1979 and again from 2000–2006 (p. 1962). They also contend that the Sistan-Baluchistan region in Iran suffered from significant drought, leading to sharp declines in the surface water levels of the Chahnimehs from 1999–2004. However, their research outcome suffers from a lack of reliable measured data, as they relied on satellite-based data, which excludes infiltration, evaporation, and snowmelt data (Roodari et al. 2021, p. 1962). Additionally, using a mix of satellite-based data and field data from the Sistan region of Iran, Sharifikia (2013), asserts that the lower section of the Helmand basin is becoming more susceptible to drought. He states, for example, that the Sistan region of Iran received 16,000 million cubic meters (MCM) of water from the Helmand River between 1990–1991, whereas the flow dramatically declined to 40–50 MCM of water from 2000–2001 due to climatic changes and severe drought (p. 2010). Similarly, Afghan researchers such as Alami and Tayfur (2022) conclude that the Helmand River Basin suffered from extreme drought from 1993–2001 and from severe drought from 2001–2004 (p. 9). More recently, Jain (2018) argued that Iran received more than 3 billion cubic meters of water in 2017 despite the drought that hit the basin during 2017–2018, a claim that has been rejected by Iranian authorities (Iran Front Page 2022). As the level of mistrust between the two riparian states has continued to grow, the water conflict between these two riparian states has also increased (Mianabadi et al. 2020).

Authors such as Akhtar and Shah (2019), Sharifikia (2013), and Loodin and Warner (2022) argue that Iran took advantage of the chaos in Afghanistan that began in the mid-1970s by constructing and expanding water infrastructure within its own territory in the downstream segment of the Helmand River Basin without alerting Afghanistan of its plans. According to Bazzi et al. (2021), Iran built several large and medium-size dams on the lower part of the basin (see also Sistan-Baluchistan Province Portal 2011). Additionally, Iran constructed four Chahnimehs (reservoirs) through which a significant portion of the Helmand River flow is now diverted. The total storage capacity of these four Chahnimehs is 2,900 million cubic meters (Dahmardeh Ghaleno et al. 2017). According to Afghan authorities interviewed for this study, these infrastructural changes in Iran are partially responsible for Iran receiving more than its fair share of Helmand River water as compared to its allotment under the 1973 treaty (Ehsan, March 17, 2022, 18:49; Mahmood, Mar 12, 2022, 01:25). However, by installing the Chahnimehs, diverting the Helmand River into these reservoirs, and then transferring the water to Zahedan city, which is located 200 km away from Helmand Fork and lies outside the Helmand basin, Hussaini and Azimi (2021a) contend that Iran has actually triggered water shortages and the desiccation of the hamouns in both Iran and Afghanistan. Mayar (2020) believes that Iran currently transfers 26 million cubic meters of water annually to Zahedan city (see also Ali 2020). As a result, the hamouns of Gowd-i-Zerah (McMahon 1906) and Puzak in Nimroz Province have dried up, which has led to a massive migration of Afghans to other parts of the country (Rasmussen 2017).

Shroder and Ahmadzai (2016), on the other hand, indicate that Iran has exacerbated the situation of water scarcity by over-pumping the lower basin's transboundary aquifers, causing cross-border groundwater depletion (p. 413). Shroder (2016) indicates that these withdrawals have lowered the water table across the border in Nimroz

Province in southwestern Afghanistan (p. 350), where the 630,000 residents of the province currently suffer from a lack of water for drinking and irrigation (Salehi 2021; Bhattacharyya et al. 2004; Kakar 2021). Nimroz residents now have no choice but to buy water from tanker owners who import water from Iran (Zurutuza 2014). Moreover, the combined decline in both surface and groundwater flows has further prompted Nimroz residents to migrate away (Kakar 2021).

10.4 Interpreting the State of Data Sharing Between Iran and Afghanistan from the Perspective of the TPB Model

10.4.1 Data and Information Sharing Regarding the Helmand River Basin

In the context of the Helmand River, except for data on measured discharge rates Afghan commissioners shared in hard copy between 2004 and 2021 with their Iranian counterparts from Khwabgah Station (Goes et al. 2015; Hearns 2015; Mohammad, February 23, 2022; 10:15), no other formal or informal channel of data sharing has been reported. Between 1980–2004, no discharge data was shared largely because Afghanistan's hydrometric stations were destroyed during the social unrest that afflicted the country. Clearly, the lack of sharing reliable data on shared waterways hampers water negotiation and cooperation among riparian states (Gerlak et al. 2011, 2014). This is one of the chief reasons why Afghanistan and Iran have not been able to develop any coordinated drought and flood mitigation strategies or water use efficiencies in the Helmand River Basin (Suryavanshi et al. 2021).

The reason data sharing has been limited is that neither Afghanistan nor Iran has implemented regular data exchange procedures. While Article 5 of Protocol # 1 to the 1973 Helmand River Treaty states that "the Afghan commissioner shall put at his disposal [referring to the Iranian commissioner] the related data registered at that station [Dehrawud Station]" and that "should the Iranian commissioner so desire, the Afghan commissioner shall put at his disposal the monthly flow records registered at that station" (Art. 5, Protocol #1), neither of these provisions has ever been implemented.

In fact, the Helmand River Treaty lacks a true mechanism for data sharing. First, it only created a one-way process (Mohammad, February 23, 2022, 10:23), as only Afghanistan is responsible for sharing data with Iran upon its request (Helmand River Treaty 1973). Iranian commissioners, on the other hand, have no obligation to share data under the treaty and have not done so voluntarily due to the lack of shared water infrastructures near the political border of Afghanistan and Iran. Secondly, under the treaty, the two riparian states are responsible for jointly building "technical structures" through which they can sustainably utilize the water resources of the river. However, as revealed by the analysis, the two and half decades of internal war in

Afghanistan, along with a lack of political will by both Iran and Afghanistan, have made it considerably difficult to construct the necessary joint structures (see Shroder and Ahmadzai 2016; Ramachandran 2022). Moreover, Dehrawud Station, the most important hydrometric station for data sharing according to the treaty, was destroyed during the social unrest in Afghanistan and was inaccessible between 2001–2020, when the Taliban controlled the lower Helmand basin region. The Afghan government was only able to re-activate the station in 2020, once the region was liberated from the Taliban.

Additionally, under the treaty, the Afghan government was responsible for constructing the Dehrawud hydrometric station. The purpose of this station is solely to assess when "a normal water year" occurs (Helmand River Treaty 1973, article 1[d]). This station, however, was not constructed until 2021 due to the ongoing conflicts between the Afghan government and the Taliban in southern and south-eastern Afghanistan, especially in Helmand Province (2001–2021) (Mohammad, February 23, 2022, 09:30).

Some of the Afghan and international experts interviewed for this study accused Iran of supporting the Taliban to disrupt security in the region so that Iran can utilize the Helmand River without interference or objection (Mahmood, Mar 12, 2022, 01:25; Kabir, Mar 30, 2022, 17:00). They contend that if the Afghan government had been able to construct Dehrawud Station earlier in the century, Iran would not have been able to use more water than its rightful share from the Helmand River (Sharif, March 13, 2022, 13:03). This claim is frequently rejected by Iranian authorities and scholars (Tahir, March 13, 2022, 21:45).

While Dehrawud Station did not exist until 2021, the Afghan Water Resources Department measured discharge rates at Khwabgah Station (Mohammad, February 23, 2022, 09:50). Those measurements were delivered to the Iranian authorities during the official meetings of the two countries' commissioners since 2004.

According to interviewee Mohammad, from 2004 until the Taliban took over Afghanistan in August 2021, commissioners from both riparian nations held 24 official meetings in both countries, but neither country has ever made any official records of these meetings publicly available (Faizee 2022). At these meetings, Afghan commissioners shared the hard copy of their annual discharge rate measurement at Khwabgah Station with their Iranian counterparts. According to interviewee Mohammad, no other type of data (e.g. hydrological, meteorological, groundwater, or hydro-morphological data) was shared by the Afghan commissioners with Iran. Although exchange through email and other online platforms (e.g. uploading data through a website, or via open data sources) is the most common means for data exchange among riparian states (Mukuyu et al. 2020), Afghanistan has only ever shared hard copies of discharge rate data with Iran.

10.4.2 *Identifying the Causes of the Lack of Data Sharing*

As established by the TPB model, the commitment to data sharing is shaped by attitude, national interest, and other "pluses and minuses." In the case of Afghanistan and Iran, the primary cause for the limited data sharing has been the historical lack of trust between the two countries (Mianabadi et al. 2020), which affects their attitude and their perceptions of national interest. According to Mianabadi et al., despite the Afghan government having shared its data with their Iranian counterparts for more than sixteen years (2004–2021), Iran still does not trust the data collected by Afghanistan (see Mianabadi et al. 2020, p. 3459; Shroder 2016, p. 356). As discussed above, government officials and scholars from the two countries offer very different opinions on the flow rates reaching Iran, as well as on the amount of water that both countries are entitled to. As a result, the two riparians have consistently disagreed over water rights and allocations for decades, including the application of the 1973 treaty.

Some of the Afghan authorities and international experts who were interviewed for this project believe that Iran has never been honest in data sharing. They argue that Iranian officials know how much water they have received during the two and half decades during which Afghanistan was mired in conflict (Goes et al. 2015), as well as over the last two decades, when the basin experienced significant tension between the government and the Taliban (Akhtar and Shah 2019). Throughout this time, Hussaini and Azimi asserted that Iran has strategically argued that its water rights have been violated by Afghanistan's upstream uses of the Helmand River (Mahmood, Mar 12, 2022, 01:30; Hussaini and Azimi 2021a, 2021b).

One reason why Iran may not be committed to regular data sharing in the basin is that it has realized that its current water rights under the treaty (820 million cubic meters of water in a normal water year) are not sufficient to meet its water needs (Mahmood, Mar 12, 2022, 01:43). Consequently, it is not in its national interest to respect the agreement. In fact, Iran's water needs have increased due to a dramatic increase in population, coupled with climatic changes and unsustainable irrigation practices in Iran (see Madani et al. 2016; Madani 2014, 2021). Iran's present water needs from the Helmand River amount to three times its treaty-based water rights (Shahab, February 27, 2022, 13:15). Thus, according to interviewee Mohammad, Iranian authorities believe that sharing data with Afghanistan, including historical data from 1975–2001, will directly threaten Iran's national security (Mohammad, Feb 23, 2022, 10:25). Thus, we argue that the scholarship of International Relations (IR) should be further explored, as a number of important factors, such as fear, threat, pride, national identity, or emotion (i.e. challenges of data sharing), are still un-explored from the discourse of transboundary water governance.

Iran and Afghanistan have also failed to cooperate on filling the data gap that developed between 1980 and 2004 due to domestic turmoil. Mustafa, an interviewee from Iran, believes that Iranian officials may now be ready to jointly work with Afghan officials on filling these gaps and mitigating the water conflict between the two nations (Mustafa, March 18, 2022, 20:00). However, Afghan authorities have repeatedly requested their Iranian counterparts to help Afghanistan fill in the missing data, and Mohamad, an interviewee from Afghanistan, reported that Iran has always rejected these requests (Mohammad, February 23, 2022, 10:16). This may be because sharing data on the Helmand River Basin is very sensitive and serves as a national security priority for both states, especially for Iran (VOA Persian 2022).

10.5 The 2013 Comprehensive Strategic Cooperation Document

To address its water needs, in 2013 Iran proposed that the two governments jointly develop and sign the "Comprehensive Strategic Cooperation Document" (CSCD). While the aim of the document initially was to strengthen bilateral cooperation through transportation and trade (Panda 2015), the former Minister of Energy and Water of Afghanistan claimed that the aim of the document was to revise the Helmand River Treaty to meet Iran's water needs in return for financial support (Osmani 2021). If the CSCD had been signed by the two countries, more water would have been allocated to Iran than what is provided for in 1973 treaty (Mayar 2020; Hussaini and Azimi 2021a), which would have decreased the current flow of the Helmand River for agriculture and domestic purposes in Afghanistan's Nimroz province. Furthermore, Hussaini and Azimi (2021b) emphasize that the new agreement would have further worsened the socio-economic situation of the residents of Nimroz and Helmand provinces, who rely on the flow of the Helmand River.

Although the CSCD was finalized by both governments and was about to be formally signed, Afghan water professionals and scholars highlighted the adverse consequences of the agreement in summits held both inside and outside Afghanistan and on Afghan television networks. As a result, they were able to convince the former president of Afghanistan not to sign the document (Osmani 2021). Given the ongoing climatic changes and drought in the basin, Mayar (2020) believes that if the Helmand River Treaty were revised in a way that meets Iran's current water need, which is three times higher than what is enabled by the current treaty (Thomas and Varzi 2015), the residents of Helmand and Nimroz provinces in Afghanistan would suffer from even worse water shortages and drought then they are experiencing now. In fact, drought has already affected the livelihoods of local Afghan farmers in the lower Helmand River Basin, resulting in massive migration from villages and farms to nearby cities in Afghanistan (Jain 2018; Kakar 2021; Ali 2021).

After the collapse of the Afghan government in August 2021, Iran established close relations with the Taliban administration, a government that has been deemed illegitimate by the majority of the world's nations (Faiez 2022). Faizee (2022) notes that since the Taliban took over, Iran has held two official meetings with the Afghan commissioners, but no agreement has been reached. Given the dramatic decline in the flow of the Helmand River due to climatic changes, coupled with an increase in population, Iran's water needs have grown significantly. To address this need, Iran has requested a meeting of the "committee of ministers" under the treaty, which includes representatives from both the Iranian and the Taliban governments. It is worth noting that the call for a meeting of the committee can happen only in an emergency situation or when the commissioners of both riparian nations are unable to reach an agreement (see article 10(c), Helmand River Treaty 1973). However, the current government of Afghanistan under Taliban administration lacks the legitimacy and technical expertise to facilitate and participate in these committee meetings. Faizee (2022) contends that even if such meetings take place, the committee will not privilege Iran with more water than its rightful share of water from the Helmand River as provided for in the 1973 treaty.

The Iranian actions are rather unique from the perspective of the theoretical literature on transboundary water management. Contrary to the expectation that upstream riparian countries tend to limit downstream riparian nations from accessing water use and water abstraction data from their section of the basin (Affeltranger et al. 2009), in the case of the Helmand River Basin, the roles are interchanged. It is the downstream riparian state, namely Iran, which has restricted Afghanistan's access to its data, a situation that was made possible by the chronic instability in Kabul. According to various authors, the Iranian government has tried to capitalize on the developments taking place in Afghanistan in many ways. Ramachandran (2017, 2022), an Indian journalist, argues that Iran supported the Taliban's efforts to disrupt the construction of Kamal Khan Dam to prevent Afghanistan from using more Helmand River water. Moreover, Aman (2016), an Iranian-born journalist, argues that when the Afghan government initiated the third phase of the dam, Iran tried to use the three million Afghan refugees in Iran as leverage to coerce the Afghan government (p. 4) to stop the construction of the dam. Shroder (2016) explains that Iranian authorities have also used undocumented Afghan immigrants and refugees in Iran as a bargaining chip in other negotiations related to the Helmand River. Similarly, Osmani (2021), former Minister of Energy and Water of Afghanistan, argues that in 2020, Iran used the refugee issue against Afghanistan, threatening to deport the three million illegal Afghans in Iran to try to force the Afghan government to revisit the Helmand River Treaty by having the Afghan government sign the CSCD.

10.6 Conclusion

Despite the importance of the exchange of reliable data and information in transboundary river basins, the Helmand River case study shows that in practice, there are factors that hamper the process of sharing between riparian nations on a transboundary watercourse. Employing a mixed study approach, this research borrowed from the TPB model to illustrate the challenges of data sharing for the transboundary Helmand River Basin shared between Afghanistan and Iran. While the TPB model is well-developed and is widely applied in various international river basins, we felt that the model has under-represented some factors that challenge the reliable data exchange mechanism between riparian nations. Thus, we developed a modified version of the model, which unravels some of the underlying challenges of data sharing between Afghanistan and Iran, the riparian nations of the Helmand River Basin. Relying on stakeholder interviews and secondary literature data, this chapter found that emotions, history and pride, national identity, and securitization are critical factors of national interests that have impeded the Helmand River Basin riparian states from cooperating on data sharing.

For example, given the drastic climate changes on the flow of the Helmand River, especially in the downstream, the demand for water use has tripled in the Sistan region due to the rapid increase in population. On the other hand, as both riparian nations have failed to specify the three points of water delivery, the water conflict has intensified between the two nations. While Afghanistan accuses Iran of the desiccation of the Hamoun wetlands, including Gowd-i-Zerah, the downstream state of Iran blames Afghanistan for blocking the flow of the Helmand River through the inauguration of Kamal Khan Dam. As water serves the national interest of both countries, we argued that TPB should be modified to uncover the current challenges of data sharing in the Helmand River Basin. It was also contended that the discourse of transboundary water governance, especially from the perspective of data sharing, should be revisited and further explored, as a number of important factors, e.g. emotions, national identity, pride, threat, or fear are under-estimated or under-explored by theories of International Relations. Therefore, our findings serve as a basis for future research on the discourse of neo-liberalism and how neo-liberal theory highlights the challenges of data sharing in a water-stressed basin, e.g. the Helmand River, and finally how the incorporation of constructivist elements into a model of neo-liberalism may lead to the sustainable utilization of a shared river through a reliable data sharing mechanism.

References

Abidi AHH (1977) Irano-Afghan dispute over the Helmand waters. Int Stud 16(3):357–378. https://doi.org/10.1177/002088177701600304

Affeltranger B (2009) Sustainability of environmental regimes: The Mekong River Commission. In: Brauch H, Spring Ú, Grin J et al (eds) Facing global environmental change, vol 4. Springer, Berlin Heidelberg, pp 593–601

Ahmed AT, El Gohary F, Tzanakakis VA, Angelakis AN (2020) Egyptian and Greek water cultures and hydro-technologies in ancient times. Sustainability 12(22):9760. https://doi.org/10.3390/su12229760

Ajzen I (1985) From intentions to actions: a theory of planned behavior. In: Kuhl J, Beckmann J (eds) Action Control. Springer, Berlin Heidelberg, pp 11–39

Akhtar F, Shah U (2019) Emerging water scarcity issues and challenges in Afghanistan. In: Ranjan A (ed.) Water issues in Himalayan South Asia. Palgrave Macmillan Singapore, Singapore, pp 1–28

Alami MM, Tayfur G (2022) Meteorological drought analysis for Helmand River Basin, Afghanistan. Teknik Dergi 33(4):12223–12242. https://doi.org/10.18400/tekderg.868595

Ali (2020) Iran's deep wells illegally siphon water from three Afghan rivers. Salaam Times. https://afghanistan.asia-news.com/en_GB/articles/cnmi_st/features/2020/09/04/feature-01. Accessed 2 June 2022

Ali (2021) Afghans recount torture in Iran in retaliation for Kamal Khan Dam opening. Salaam Times. https://afghanistan.asia-news.com/en_GB/articles/cnmi_st/features/2021/05/13/feature-02. Accessed 7 June 2022

Aman F (2016) Water dispute escalating between Iran and Afghanistan. Atlantic Council. https://www.atlanticcouncil.org/wp-content/uploads/2016/09/Water_Dispute_Escalating_between_Iran_and_Afghanistan_web_0830.pdf. Accessed 7 June 2022

Amini A, Ghoreishi SZ, Mianabadi H (2021) Understanding 1973 the Helmand Treaty by invoking rules of interpretation according to Vienna Convention 1969. J Water Manag Irrig 11(2):249–273. https://doi.org/10.22059/JWIM.2021.324894.876

Amiri F, Rahdari V, Maleki Najafabadi S, Pradhan B, Tabatabaei, T (2013) Multi-temporal landsat images based on eco-environmental change analysis in and around Chah Nimeh Reservoir, Balochestan (Iran) Environ Earth Sci 72(3):801–809. https://doi.org/10.1007/s12665-013-3004-9

Armitage D, de Loë RC, Morris M et al (2015) Science–policy processes for transboundary water governance. Ambio 44(5):353–366. https://doi.org/10.1007/s13280-015-0644-x

Askari N, Bashardost N (2021) Afghanistan no longer relinquishes free water: Ghani. Pajhwok. https://pajhwok.com/2021/03/24/afghanistan-no-longer-relinquish-free-water-ghani. Accessed 10 April 2022

Barua A, Vij S, Zulfiqur RM (2018) Powering or sharing water in the Brahmaputra River basin. Int J Water Resour Dev 34(5):829–843. https://doi.org/10.1080/07900627.2017.1403892

Bazzi H, Ebrahimi H, Aminnejad B (2021) A comprehensive statistical analysis of evaporation rates under climate change in southern Iran using WEAP (case study: Chahnimeh Reservoirs of Sistan Plain). Ain Shams Eng J 12(2):1339–1352. https://doi.org/10.1016/j.asej.2020.08.030

Beniston M, Stoffel M, Harding R et al (2012) Obstacles to data access for research related to climate and water: implications for science and EU policy-making. Environ Sci Policy 17:41–48. https://doi.org/10.1016/j.envsci.2011.12.002

Bergner RM (2011) What is behavior? And so what? New Ideas Psychol 29(2):147–155. https://doi.org/10.1016/j.newideapsych.2010.08.001

Bhattacharyya K, Azizi PM, Shobair SS, Mohsini MY (2004) Drought impacts and potential for their mitigation in Southern and Western Afghanistan. International Water Management Institute. https://www.iwmi.cgiar.org/Publications/Working_Papers/working/WOR91.pdf. Accessed 21 Dec 2022

Bidabadi FS, Afshari L (2020) Human right to water in the Helmand Basin: setting a path for the conflict settlement between Afghanistan and Iran. Utrecht Law Rev 16(2):150–162. https://doi.org/10.36633/ulr.566

Boelens RA (2015) Water, power and identity. The cultural politics of water in the Andes. Routledge, London

Botterweg T, Rodda DW (1999) Danube River basin: progress with the environmental programme. Water Sci Technol 40(10):1–8. https://doi.org/10.1016/S0273-1223(99)00666-6

Brady LP (2017) The politics of negotiation: America's dealings with allies, adversaries, and friends. The University of North Carolina Press, Chapel Hill & London

Burton M, Molden D (2005) Making sound decisions: information needs for Basin Water Management In: Svendsen M (ed.) Irrigation and river basin management: options for governance and institutions. Cabi Publishing, Cambridge, pp 51–74

Calhoun CJ (1997) Nationalism. University of Minnesota Press, Minneapolis

Chenoweth J, Feitelson E (2001) Analysis of factors influencing data and information exchange in international river basins: can such exchanges be used to build confidence in cooperative management? Water Int 26(4):499–512. https://doi.org/10.1080/02508060108686951

Cherry K (2022) How can our attitudes change and influence behaviors? Verywell Mind. https://www.verywellmind.com/attitudes-how-they-form-change-shape-behavior-2795897 Accessed

19 Dec 2022
Climate Diplomacy (2016) Transboundary water disputes between Afghanistan and Iran. Climate. https://climate-diplomacy.org/case-studies/transboundary-water-disputes-between-afghanistan-and-iran. Accessed 3 Oct 2022
Coelho VA, Sousa V, Marchante M (2016) Social and emotional competencies evaluation questionnaire—Teacher's version: validation of a short form. Psychol Rep 119(1):221–236.
Dabiri MR (2011) بررسی حقوقی و تاریخی رودخانه هیرمند [Legal and historical study of Helmand river (Helmand)]. Diplomacy Irani. http://irdiplomacy.ir/fa/news/11868. Accessed 11 April 2022
Dahmardeh Ghaleno MR, Sadoddin A, Sheikh V, Sabouhi Sabouni M (2017) Optimal utilization of the Chahnimeh water reservoirs in Sistan region of Iran using goal programming method. Ecopersia 5(1):1641–1654
Dominguez FJ, Lowry BL, Webb CE (1951) The Helmand River Basin Dispute. https://aquapedia.waterdiplomacy.org/wiki/index.php?title=The_Helmand_River_Basin_Dispute. Accessed 14 March 2022
Dormido H (2019) These countries are the most at risk from a water crisis. Bloomberg. https://www.bloomberg.com/graphics/2019-countries-facing-water-crisis. Accessed 15 March 2022
Eckstein GE (2005) Protecting a hidden treasure: The U.N International Law Commission and the international law of transboundary ground water resources. Sustain Dev Law Policy 5(1):5–12
Edensor T (2009) National spatialities. In: Kitchin R, Thrift N (eds) International encyclopedia of human geography. Elsevier Science, India, pp 242–247
Faiez R (2022) Taliban acknowledge struggle for recognition of Afghan govt. AP NEWS. https://apnews.com/article/afghanistan-poverty-taliban-food-crisis-4741a4fedd27ab986720f06a619d2648. Accessed 3 Oct 2022
Faizee M (2022) The emerging dynamics for conflict and cooperation between Iran and the Taliban over the Helmand River. Water, peace, and Security. https://waterpeacesecurity.org/info/blog-09-09-2022-The-emerging-dynamics-for-conflict-and-cooperation-between-Iran-and-the-Taliban-over-the-Helmand-River. Accessed 28 Sept 2022
Foltz RC (2002) Iran's water crisis: cultural, political, and ethical dimensions. J Agric Environ Ethics 15(4):357–380. https://doi.org/10.1023/A:1021268621490
Frederickson BL, Joiner T (2002) Positive emotions trigger upward spirals towards emotional wellbeing. Psychol Sci 13(2):172–175. https://doi.org/10.1111/1467-9280.00431
Furlong K, Petter Gleditsch N, Hegre H (2006) Geographic opportunity and neo-Malthusian willingness: boundaries, shared rivers, and conflict. International Interactions 32(1):79–108. https://doi.org/10.1080/03050620600596421
Gawronski B (2007) Attitudes can be measured! But what is an attitude? Soc Cogn 25(5):573–581. https://doi.org/10.1521/soco.2007.25.5.573
Gendron M (2010) Defining emotion: a brief history. Emot Rev 2(4):371–372. https://doi.org/10.1177/175407391037466
Gerlak AK, Grant KA (2009) The correlates of cooperative institutions for international rivers. In: Volgy TJ, Sabic Z, Roter P, Gerlak AK (eds) Mapping the new world order. Wiley-Blackwell Publishers, New York, pp 114–147
Gerlak AK, Lautze J, Giordano M (2011) Water resources data and information exchange in transboundary water treaties. Int Environ Agreements Polit Law Econ 11(2):179–199. https://doi.org/10.1007/s10784-010-9144-4
Gerlak AK, Lautze J, Giordano M (2014) Greater exchange, greater ambiguity: water resources data and information exchange in transboundary water treaties. In: Grafton RQ, Wyroll P, White C, Allendes D (eds) Global water: issues and insights. Australian University Press, Canberra, pp 57–64
Gharesifard M (2015) Mapping the behavioural determinants of ICT-based citizen participation in water management; case studies of sharing personally-collected weather data via web-platforms in the Netherlands, UK and Italy. UNESCO. https://ihedelftrepository.contentdm.oclc.org/digital/collection/masters2/id/94090. Accessed 28 Sept 2022.

Gharesifard M, Wehn U (2015) Participation in citizen science: drivers and barriers for sharing personally-collected weather data via web-platforms. In: Paper Presented at the 8th International Conference on ICT, Society and Human Beings, Las Palmas de Gran Canaria, Spain. https://www.researchgate.net/profile/Mohammad-Gharesifard/publication/277009826_Drivers_and_barriers_for_sharing_personally-collected_weather_data_via_web-platforms/links/5ffffda445851553a0419119/Drivers-and-barriers-for-sharing-personally-collected-weather-data-via-web-platforms.pdf. Accessed 22 Sept 2022

Gharesifard M, Wehn U (2016) To share or not to share: drivers and barriers for sharing data via online amateur weather networks. J Hydrol 535:181–190. https://doi.org/10.1016/j.jhydrol.2016.01.036

Gleditsch NP, Furlong K, Hegre H et al (2006) Conflicts over shared rivers: resource scarcity or fuzzy boundaries? Polit Geogr 25(4):361–382. https://doi.org/10.1016/j.polgeo.2006.02.004

Goes BJM, Howarth SE, Wardlaw RB et al (2015) Integrated water resources management in an insecure river basin: a case study of Helmand River Basin Afghanistan. Int J Water Resour Dev 32(1):3–25. https://doi.org/10.1080/07900627.2015.1012661

Gul A (2021) Afghan leader demands Iranian oil in exchange for river water. VOA News, https://www.voanews.com/a/south-central-asia_afghan-leader-demands-iranian-oil-exchange-river-water/6203699.html. Accessed 10 April 2022

Hearns G (2015) The Helmand River and the Afghan-Iranian Treaty of 1973. International Water Law Project Blog. https://www.internationalwaterlaw.org/blog/2015/07/23/the-helmand-river-and-the-afghan-iranian-treaty-of-1973. Accessed 20 April 2022

Helmand River Treaty (1973) The Afghan Iranian Helmand River Water Treaty. International Water Law. https://www.internationalwaterlaw.org/documents/regionaldocs/1973_Helmand_River_Water_Treaty-Afghanistan-Iran.pdf. Retrieved 15 April 2022

Holmwood J (2017) The university, democracy and the public sphere. Br J Sociol Educ 38(7):927–942. https://doi.org/10.1080/01425692.2016.1220286

Hume D (1896) A treatise of human nature. Clarendon Press, Oxford

Hussaini SA, Azimi AB (2021a) Statute for illegal water rights, Helmand River Treaty is the only water treaty. 8am Newspaper Afghanistan. https://8am.media/articles-of-association-for-unfounded-payments-only-the-helmand-treaty-is-the-basis. Accessed 27 March 2022

Hussaini SA, Azimi AB (2021b) Where is the water of Hamuns. 8am Newspaper Afghanistan. https://8am.media/where-is-the-water-of-hamuns. Accessed 28 March 2022

Hutchings P, Cooper S, Butterworth J et al (2022) Water and emotion: testing a new approach for monitoring water security among afar pastoralists in Ethiopia. Front Clim 3(75388):1–12. https://doi.org/10.3389/fclim.2021.753888

Hutchison E (2016) Affective communities in world politics: collective emotions after trauma. Cambridge University Press, Cambridge

Iceland C, Otto B (2017) What does water have to do with national security? World Resources Institute. https://www.wri.org/insights/what-does-water-have-do-national-security. Accessed 1 June 2022

Iqbal MW, Donjadee S, Kwanyuen B, et al. (2018) Farmers' perceptions of and adaptations to drought in Herat Province, Afghanistan. J Mt Sci 15(8):1741–1756. https://doi.org/10.1007/s11629-017-4750-z

Iran Front Page (2022) Official: Kabul must prove respect for Iran's Helmand water share in action. Iran Front Page. From https://ifpnews.com/iran-afghanistan-helmand-water-share. Accessed 3 Oct 2022

Jain R (2018) In parched Afghanistan, drought sharpens water dispute with Iran. Reuters. https://www.reuters.com/article/us-afghanistan-iran-water-idUSKBN1K702H. Accessed 9 June 2022

Kakar AZ (2021) Nimroz residents grumble lack of clean drinking water. Pajhwok. https://pajhwok.com/2021/12/12/nimroz-residents-grumble-lack-of-clean-drinking-water. Accessed 2 June 2022

Kakar AZ, Sayedi N (2022) Kamal Khan dam water released, but where is it flowing? Pajhwok. https://pajhwok.com/2022/01/20/kamal-khan-dam-released-but-where-is-it-flowing. Accessed

9 April 2022

Keohane RO, Nye JS (1973) Power and interdependence. Survival 15(4):158–165

Kerr J, Privatdozen RW, Moretti M (2005) ABC of behavior change. Elsevier, Philadelphia

Kliot N, Shmueli D (2001) Development of institutional frameworks for the management of transboundary water resources. Int J Glob Environ Issues 1(3/4):306–328. https://doi.org/10.1504/ijgenvi.2001.000982

Kliot N, Shmueli D, Shamir U (2001) Institutions for management of transboundary water resources: their nature, characteristics and shortcomings. Water Policy 3(3):229–255. https://doi.org/10.1016/s1366-7017(01)00008-3

LaMorte WW (2022) Behavioral change models. The Theory of Planned Behavior. Boston University. https://sphweb.bumc.bu.edu/otlt/mph-modules/sb/behavioralchangetheories/BehavioralChangeTheories3.html. Accessed 19 Dec 2022

Land Links (2018) Afghanistan. LandLinks. https://land-links.org/country-profile/afghanistan/#water. Accessed 26 May 2022

Lane AM (2020) The securitization of transboundary water resources: a discourse analysis of Iranian-Afghan relations in the Helmand River Basin. Thesis, Georgetown University. https://repository.library.georgetown.edu/handle/10822/1060360. Accessed 22 Feb 2022

Liu Z (2013) The concept of national interests. In: Yang J (ed) China's diplomacy—Theory and practice. World Century Publishing Corporation, Hackensack, pp 121–189

Loodin N, Warner J (2022) A review of hydro-hegemonic dynamics on the transboundary Harirud River Basin: 2001–present. Water 14(21):3442. https://doi.org/10.3390/w14213442

Loodin N, Wolf AT (2022) Will Islamic water management principles be included if the Helmand River treaty is revisited? Water 14(1):67. https://doi.org/10.3390/w14010067

Madani K (2014) Water management in Iran: what is causing the looming crisis? J Environ Stud Sci 4(4):315–328. https://doi.org/10.1007/s13412-014-0182-z

Madani K (2021) Explainer: Iran's "water bankruptcy". The Iran Primer. https://iranprimer.usip.org/blog/2021/dec/05/explainer-irans-water-bankruptcy. Accessed 24 April 2022

Madani K, Aghakouchak A, Mirchi A (2016) Iran's socio-economic drought: challenges of a water-bankrupt nation. Iran Stud 49(6):997–1016. https://doi.org/10.1080/00210862.2016.1259286

Mayar A (2020) Afghanistan-Iranian strategic Accord-Water. Tolonews. https://tolonews.com/pa/opinion/ د افغانستان-ایران په ستراتیژیک تړون کې د اوبو برخه Accessed 27 March 2022

Mayar A (2022) The climate change crisis in Afghanistan: the catastrophe worsens—what hope for action? Afghanistan Analysts Network. https://www.afghanistan-analysts.org/en/reports/economy-development-environment/the-climate-change-crisis-in-afghanistan-the-catastrophe-worsens-what-hope-for-action. Accessed 16 June 2022

McMahon H (1906) Recent survey and exploration in Seistan. Geogr J 28(3):209–228

Mehdi SZ (2021) Hope sparks in Iran to settle water dispute with Afghanistan. Anadolu Agency https://www.aa.com.tr/en/middle-east/hope-sparks-in-iran-to-settle-water-dispute-with-afghanistan/2379941. Accessed 9 April 2022

Mianabadi A, Davary K, Mianabadi H et al (2020) International environmental conflict management in transboundary river basins. Water Resour Manage 34(11):3445–3464. https://doi.org/10.1007/s11269-020-02576-7

Mianabadi H, Alioghli S, Morid S (2021) Quantitative evaluation of 'no-harm' rule in international transboundary water law in the Helmand River Basin. J Hydrol 599(126368). https://doi.org/10.1016/j.jhydrol.2021.126368

Milani MM (2018) The making of Iran's Islamic revolution: from monarchy to Islamic republic. Routledge, New York & London

Mukuyu P, Lautze J, Rieu-Clarke A et al (2020) The devil's in the details: data exchange in transboundary waters. Water Int 45(7–8):884–900. https://doi.org/10.1080/02508060.2020.1850026

Myers DG, Smith SM (2012) Exploring social psychology. McGraw-Hill, New York

Nagheeby M, Rieu-Clarke A (2020) Water diplomacy in the Helmand River Basin: exploring the obstacles to cooperation within the shadow of anarchy. In: Kittikhoun A, Schmeier (eds.) River basin organizations in water diplomacy. Routledge, London, pp 201–221

Najafi A, Vatanfada J (2013) Transboundary water management improvements, the way forward in the Middle East; case study: transboundary water management of Iran and neighbors. Geogr Q 8(4):135–155

Ong A (2006) Neoliberalism as exception: mutations in citizenship and sovereignty. Duke University Press, Durham, New York, and London

Osmani AA (2021) Afghanistan's water resources are in the grip of Iran's reverse cycle. 8am Newspaper Afghanistan. https://8am.media/afghanistans-water-resources-are-in-the-grip-of-irans-reverse-cycle. Accessed 27 March 2022

Panda A (2015) Iran, Afghanistan approach strategic cooperation pact. The Diplomat. https://thediplomat.com/2015/01/iran-afghanistan-approach-strategic-cooperation-pact. Accessed 9 June 2022

Plengsaeng B, Wehn U, van der Zaag P (2014) Data-sharing bottlenecks in transboundary integrated water resources management: a case study of the Mekong River Commission's procedures for data sharing in the Thai context. Water International 39(7):933–951. https://doi.org/10.1080/02508060.2015.981783

Pooreh S, Nodeh ZH (2015) Impact of education based on theory of planned behavior: an investigation into hypertension-preventive self-care behaviors in Iranian girl adolescent. Iran J Public Health 44(6):839–847

Porta EL, Wolf AT (2021) Intrinsic and spiritual dimensions of water at the local scale, and the disconnect with international institutions. Sustainability 13(16):8948. https://doi.org/10.3390/su13168948

Prasai S (2013) Beyond South Asian geopolitics: the everyday forms of state ineffectiveness in water cooperation. Asian J J Transp Infrastruct 17(1):41–54

Price G, Alam R, Hasan S, et al. (2014) Attitudes to water in South Asia. Chatam House. https://www.chathamhouse.org/sites/default/files/field/field_document/20140627WaterSouthAsia.pdf. Accessed 9 June 2022

Ramachandran S (2017) Unquiet flows the Helmand. Asia Times. https://asiatimes.com/2017/05/unquiet-flows-helmand. Accessed 23 April 2022

Ramachandran S (2022) Afghanistan-Iran disquiet over the Helmand River. The Diplomat. https://thediplomat.com/2022/02/afghanistan-iran-disquiet-over-the-helmand-river. Accessed 9 April 2022

Rasmussen SE (2017) On the edge of Afghanistan. Foreign Policy. https://foreignpolicy.com/2017/09/12/on-the-edge-of-afghanistan-taliban-iran-war-united-states. Accessed 17 June 2022

Roodari A, Hrachowitz M, Hassanpour F et al (2021) Signatures of human intervention–or not? Downstream intensification of hydrological drought along a large Central Asian river: the individual roles of climate variability and land use change. Hydrol Earth Syst Sci 25(4):1943–1967. https://doi.org/10.5194/hess-25-1943-2021

Sadat SH, Sayed N (2020) Afghanistan and Iran: from water treaty to water dispute. The Interpreter. https://www.lowyinstitute.org/the-interpreter/afghanistan-and-iran-water-treaty-water-dispute Accessed 8 June 2022

Sadoff C, Grey D (2002) Beyond the river: the benefits of cooperation on international rivers. Water Policy 4(5):389–403. https://doi.org/10.1016/s1366-7017(02)00035-1

Saif SK (2021) Decades in making, Afghan mega dam opens on Iran border. Anadolu Agency. https://www.aa.com.tr/en/asia-pacific/decades-in-making-afghan-mega-dam-opens-on-iran-border/2186632. Accessed 30 March 2022

Salehi NA (2021) Nimruz faces water crisis despite Kamal Khan Dam. TOLO News. https://tolonews.com/afghanistan-172740 Accessed 2 June 2022

Sandstrom E, Jägerskog A, Oestigaard T (2018) Land and hydropolitics in the Nile River Basin: challenges and new investments. Routledge, London and New York

Sarfaraz MU, Hall DM, Rotman RM (2022) Data sharing in transboundary water management. Front Water 4(982605):1–7. https://doi.org/10.3389/frwa.2022.982605
Schmeier S, Gerlak AK, Blumstein S (2016) Clearing the muddy waters of shared watercourses governance: conceptualizing international River Basin Organizations. Int Environ Agreements Polit Law Econ 16(4):597–619
Seide WM (2010) On Nile, Ethiopia goes soft to get tough. https://www.academia.edu/24085851/On_Nile_Ethiopia_Goes_Soft_to_Get_Tough. Accessed 11 April 2022
Seide WM (2016) Hydro-mentality over the Nile. Horn Affairs. https://hornaffairs.com/2016/04/03/egypt-ethiopia-hydro-mentality-nile-river. Accessed 12 April 2022
Seide WM (2018) Churning waters: strategic shifts in the Nile Basin. Institute of Development Studies. https://www.academia.edu/4859806/Churning_Waters_Strategic_Shifts_in_the_Nile_Basin. Accessed 11 April 2022
Sharifikia M (2013) Environmental challenges and drought hazard assessment of Hamoun Desert Lake in Sistan region, Iran, based on the time series of satellite imagery. Nat Hazards 65(1):201–217
Sheeran P, Webb TL (2016) The intention–behavior gap. Soc Pers Psychol Compass 10(9):503–518. https://doi.org/10.1111/spc3.12265
Shirdeli A (2014) Hydropolitics and hydrology issues in Hirmand/Helmand International River Basin. Management Science Letters 4(4):807–812. https://doi.org/10.5267/j.msl.2014.2.002
Shroder JF (2014) Natural resources in Afghanistan: geographic and geologic perspectives on centuries of conflict. Elsevier, San Diego and Waltham
Shroder J (2016) Hydrogeography (drainage basins and rivers) of Afghanistan and neighboring countries. Transboundary Water Resources in Afghanistan, 23–40. https://doi.org/10.1016/b978-0-12-801886-6.00002-1
Shroder J, Ahmadzai SJ (2016) Water in Islam. In: Transboundary Water Resources in Afghanistan. Elsevier, Amsterdam
Sidky H (2007) War, changing patterns of warfare, state collapse, and transnational violence in Afghanistan: 1978–2001. Mod Asian Stud 41(4):849–888. https://doi.org/10.1017/S0026749X0600254X
Singh VP (2017) Challenges in meeting water security and resilience. Water Int 42(4):349–359. https://doi.org/10.1080/02508060.2017.1327234
Singh VP, Khedun CP, Mishra AK (2014) Water, environment, energy, and population growth: implications for water sustainability under climate change. J Hydrologic Energy 19(4):667–673
Sistan-Baluchistan Province Portal (2011) Dam construction: government's commitment for the development of Sistan-Baluchistan. Sistan-Baluchistan Province Portal. https://www.sbportal.ir/fa/news/3969/%D8%B3%D8%AF-%D8%B3%D8%A7%D8%B2%DB%8C-%D8%AA%D8%AC%D9%84%DB%8C-%D8%A7%D8%B1%D8%A7%D8%AF%D9%87-%D8%AF%D9%88%D9%84%D8%AA-%D8%A8%D8%B1%D8%A7%DB%8C-%D8%AA%D9%88%D8%B3%D8%B9%D9%87-%D8%B3%DB%8C%D8%B3%D8%AA%D8%A7%D9%86-%D9%88-%D8%A8%D9%84%D9%88%DA%86%D8%B3%D8%AA%D8%A7%D9%86. Accessed 17 April 2022
Slavnic Z (2017) Research and data-sharing policy in Sweden–neoliberal courses, forces and discourses. Prometheus 35(4):249–266. https://doi.org/10.1080/08109028.2018.1499542
Soar Earth (2022) Heavy rains won't save desert oasis. https://about.soar.earth/press-releases/heavy-rains-wont-save-desert-oasis-new-maps-show. Accessed 30 Nov 2022
Smith AD (1991) National identity. University of Nevada Press, Reno
Smith T (1998) The practice of pride. Soc Philos Policy 15(1):71–90. https://doi.org/10.1017/S0265052500003071
Spandyar YA (2020) How was the Sistan irrigation system destroyed? 8am Newspaper Afghanistan. https://8am.media/how-was-the-sistan-irrigation-system-destroyed. Accessed 30 March 2022
Storey D (2009) Political geography. In: Thrift N (ed) Kitchin R. International encyclopedia of human geography, Elsevier Science, India pp, pp 243–253

Sultana F (2011) Suffering for water, suffering from water: emotional geographies of resource access, control and conflict. Geoforum 42(2):163–172. https://doi.org/10.1016/j.geoforum.2010.12.002

Suryavanshi S, Joshi N, Maurya HK, et al. (2021) Understanding precipitation characteristics of Afghanistan at provincial scale (preprint). https://doi.org/10.21203/rs.3.rs-365073/v1. Accessed 30 March 2022

Susskind L, Islam S (2012) Water diplomacy: creating value and building trust in transboundary water negotiations. Science & Diplomacy. http://www.sciencediplomacy.org/perspective/2012/water-diplomacy. Accessed 30 March 2022

Tadros N (1996) Shrinking water resources: the national security issue of this century. Northwest J Int Law Bus 17(1):1091–1131

Thies CG (2004) Are two theories better than one? A constructivist model of the neorealist-neoliberal debate. Int Polit Sci Rev 25(2):159–183

Thomas V, Varzi MM (2015) A legal license for an ecological disaster: the inadequacies of the 1973 Helmand/Hirmand water treaty for sustainable transboundary water resources development. Int J Water Resour Dev 31(4):499–518. https://doi.org/10.1080/07900627.2014.1003346

Thomas V, Azizi MA, Behzad K (2016) Developing transboundary water resources: what perspectives for cooperation between Afghanistan, Iran and Pakistan? Afghanistan Research and Evaluation Unit Organization. https://areu.org.af/publication/1607. Accessed 15 April 2022

Thu HN, Wehn U (2016) Data sharing in international transboundary contexts: The Vietnamese perspective on data sharing in the lower Mekong Basin. J Hydrol 536:351–364. https://doi.org/10.1016/j.jhydrol.2016.02.035

Timmerman JG, Langaas S (eds) (2004) Environmental information in European transboundary water management. IWA Publishing, London

Timmerman JG, Langaas S (2005) Water information: what is it good for? The use of information in transboundary water management. Reg Environ Change 5(4):177–187. https://doi.org/10.1007/s10113-004-0087-6

Tolia-Kelly DP (2009) Hybridity. In: Kitchin R, Thrift N (eds) International encyclopedia of human geography. Elsevier Science, India, pp 258–263

Toset HPW, Gleditsch NP, Hegre H (2000) Shared rivers and interstate conflict. Polit Geogr 19(8):971–996

UNEP (2006) History of environmental change in the Sistan basin based on satellite image analysis: 1976–2005. United Nations Environment Programme. https://stg-wedocs.unep.org/bitstream/handle/20.500.11822/7690/-History_of_environmental_change_in_the_Sistan_Basin_Based_on_satellite_image_analysis_1976_%E2%80%93_2005-2006history_of_environmental_change.pdf.pdf?sequence=3. Accessed 11 April 2022

van Beek E, Bozorgy B, Vekerdy Z et al (2008) Limits to agricultural growth in the Sistan Closed Inland Delta, Iran. Irrig Drainage Syst 22(2):131–143. https://doi.org/10.1007/s10795-008-9045-7

VOA (2022) Water is considered as our national security. https://ir.voanews.com/a/qalibaf-water-national-security/6565429.html. Accessed 11 May 2022

Water Resources Department (WRD-MEW) (2022) Hydrograph of Khwabgah Station between 1973–1979 and 2010–2020, retrieved from the Ministry of Energy and Water of Afghanistan-Water Resources Department (unpublished document)

Wehn U (2000) Access to spatial data—what determines the willingness of organisations to share it? GSDI 4 Conference, Cape Town, South Africa, March 13–15

Wehn U (2001) Using the theory of planned behavior to explain spatial data sharing. ESF-NSF Workshop on Access and Participatory Approaches in Using Geographic Information. Spoleto, Italy, December 5–9

Wehn U (2003a) In search of rigorous models for policy-oriented research: a behavioral approach to spatial data sharing. J Urban Reg Inf Syst Assoc 15(1):19–28

Wehn U (2003b) Mapping the determinants of spatial data sharing, 1st edn. Routledge, London

Wester P, Rap E, Vargas-Velázquez S (2009) The hydraulic mission and the Mexican hydrocracy: regulating and reforming the flows of water and power. Water Altern 2(3):395–415

Whitney JW (2006) Geology, water, and wind in the Lower Helmand Basin, Southern Afghanistan. USGS Publications Warehouse. https://pubs.usgs.gov/sir/2006/5182. Accessed 20 Feb 2022

Wilby RL, Dessai S (2010) Robust adaptation to climate change. Weather 65(7):180–185. https://doi.org/10.1002/wea.543

Wolf AT, Stahl K, Macomber MF (2003) Conflict and cooperation within international river basins: the importance of institutional capacity. Water Resour Update 125(1):31–40

World Bank (2018) Strengthening hydromet and early warning services in Afghanistan. https://elibrary.worldbank.org/doi/abs/10.1596/31059. Accessed 20 Feb 2022

Zahid N (2017) Iran criticism of Afghan dam projects draws rebuke from Kabul officials. VOA News. From https://www.voanews.com/a/iran-afghan-dam-projects/3930411.html. Accessed 9 April 2022

Zurutuza K (2014) Afghans living under shadow of Iranian guns. Conflict. Al Jazeera. https://www.aljazeera.com/features/2014/10/15/afghans-living-under-shadow-of-iranian-guns. Accessed 9 June 2022

Part IV
New Perspectives on Transboundary Waters

Chapter 11
Discourses and Hydropolitics: Stock-Taking and Future Directions

Fatine Ezbakhe

Abstract Discourses can constitute a useful analytical tool for hydropolitics. By looking at discourses, i.e. the ideas and concepts we produce, reproduce, and transform into a particular bundle of practices, we can unpack how actors define the problems associated with water resources, the solutions considered, and the power relations underlying them. However, discourse analysis remains a relatively fuzzy approach, often operating as a "black box", which hinders its systematic and structured use in hydropolitical studies. This chapter takes stock of how discourse is understood and analyzed in hydropolitics and identifies the gaps that may serve as future directions for strengthening the discursive approach in transboundary water studies. It illustrates its claims with examples from the Nile, the Jordan, and the Mekong basins.

Keywords Hydropolitics · Discourse · Discourse analysis · Argumentative discourse analysis · Transboundary waters

11.1 Introduction

"The strength of the concept of hydropolitics, its inter-disciplinary conceptual heterogeneity, is also its weakness". This is how Grandi (2020) encapsulates the conceptual fuzziness in hydropolitics research. With its roots only dating back to the late 1970s, it is no wonder there is much conceptual fuzziness surrounding hydropolitics. The term 'hydropolitics' was first coined by Waterbury in his book *Hydropolitics of the Nile Valley*, in which he asked "how can sovereign states, pursuing national self-interest and those policies that seemingly ensure a regime's survival, cope with the challenge of bi- or multinational coordination in the use of a common resource?" (Waterbury 1979:3). Waterbury thus implicitly defined hydropolitics as the study of inter-state politics regarding the management of shared water resources. Over the

F. Ezbakhe (✉)
Institute for Environmental Sciences (ISE), UNESCO Chair in Hydropolitics, Geneva Water Hub, University of Geneva, Geneva, Switzerland
e-mail: fatine.ezbakhe@gmail.com

K. Szálkai and M. Szalai (eds.), *Theorizing Transboundary Waters in International Relations*, Springer Water, https://doi.org/10.1007/978-3-031-43376-4_11

years, scholars have put forward a series of other understandings of hydropolitics, but a quick look at the evolution of the definitions of hydropolitics (Geneva Water Hub 2021) reveals an underlying theme: hydropolitics is, in essence, the study of how state and non-state actors interact over water resources.

This blurry but blossoming research arena draws in concepts from other well-established disciplines, including political sciences, environmental sciences, and law. However, even if hydropolitics scholarship engages perspectives from critical geography, political economy, historical institutionalism, and environmental justice, among others, approaches from International Relations (IR) theories remain prevalent. Furlong (2006) called out this implicit IR theorizing in her 'hidden theories, troubled waters' piece, explaining that this implicit IR approach obfuscates the daily realities of what water cooperation and conflict between states means for those dependent on the waters in question. In her view, the tacit application of IR theories in hydropolitics can lead to a mis-theorization of hegemony, mainly in neglecting the role of ideas in conditioning the behavior of actors in world politics.

As a result, hydropolitics scholarship has also taken what is known as a discursive turn, where the analytical attention has shifted to the social construction of water politics. As Julien (2012) puts it, hydropolitics is what societies make of it, since dynamics over water resources are not only shaped by power rations and economic calculations but also by culture, history, and ideology. Hydropolitics scholars have increasingly emphasized the role of discourses (Sneddon and Fox 2006; Hussein 2017; Nagheeby and Warner 2018; Bréthaut et al. 2022; amongst others), acknowledging that transboundary waters are governed through material constructions of infrastructures (e.g., dams, canals, and irrigation networks) as much as the discursive construction of frameworks within which problems and policy responses are defined and negotiated.

This discursive perspective, however, is still relatively blurry in hydropolitics scholarship, just like it is in constructivism IR theories. While IR scholars working from a constructivist perspective emphasize the social constructs (ideas, norms, identities) underlying the international system and consider discourse an entryway to understanding how these social constructs shape social interactions, they provide different conceptualizations of discourse. This is not surprising, as discursive approaches take their intellectual roots in philosophy, linguistics, and psychology, and they lack explicitly articulated theoretical and methodological groundings. Such conceptual vagueness, however, may hinder the broad recognition of the added value of discourses and their analysis in IR in general and hydropolitics in particular.

This chapter addresses this conceptual and methodological vagueness by examining the concept of discourse and its analysis in hydropolitics. Clarifying what discourse is and how it can be analyzed to capture transboundary water interactions may bring its added value in hydropolitics research to the forefront. To this end, the chapter first takes stock of what is known about discourses and discourse analysis in hydropolitics, then it presents future directions for strengthening a systematic discursive approach in this field.

11.2 Starting Point: The Concept of Discourse and Its Analysis

When discussing discourses in hydropolitics, the starting point is acknowledging that the concept of discourse lacks a clear-cut definition. While some IR constructivists see discourse as language (Diez 1999; Fierke 2002), others see it as language *and* practice. For instance, when discussing the concept of discourse in critical geopolitics, Müller (2008) calls for veering away from the narrow, purely linguistic conception of discourse by including social practices, that is, the mechanisms by which social action unfolds. He uses the example of the bricklayer to illustrate the claim that discourse is language and practice: while the first act (asking for a brick) is linguistic, the second (adding the brick to the wall) is extra-linguistic, and the two acts together constitute the totality of building a wall.

Even when understood as both language and practice, the concept of discourse remains vague and is seldom explicitly defined. This lack of definition is clear in the most recent scholarship on discursive hydropolitics. For instance, when examining the legitimacy of nationalist discourses surrounding transboundary water interactions in the Amu Darya, Tigris-Euphrates, and Nile transboundary river basins, Allouche (2020) places discourses at the same level as norms and ideas, without explicitly specifying his understanding of the term. Similarly, Saklani et al. (2020), in their study of the securitization of the water discourse in the Ganges–Brahmaputra-Meghna river basin, do not clarify the meaning behind the notion of water discourse. Other scholars describe their interpretation of discourse inexplicitly. Hussein (2017), when examining the hydropolitics along the Yarmouk river, adopts Dryzek's (1997) relatively vague definition of discourse as "a shared way of apprehending the world". Later, Hussein et al. (2022), in identifying the tools used by state elites to justify and legitimize large-scale hydraulic infrastructure in the Euphrates-Tigris and Nile river basins, vaguely defines discourses when referring to critical discourse analysis as a method focusing on how language is constructed and used to shape norms and behavior.

Moreover, scholars of hydropolitics sometimes link the concept of discourse with those of narratives and imaginaries. For instance, when analyzing the water scarcity discourse in Jordan, Hussein (2016) relies on the notion of narratives and subnarratives to understand the 'discursive framing' of policy issues related to water scarcity in Jordan. In his words, "actors use subnarratives and narratives when trying to make sense of a discourse", as narratives constitute "stories with a beginning, middle, and end, driving therefore towards a solution" (p.11). Grandi (2021) also relies on narratives when engaging with depictions of the Nile in Egypt and Ethiopia, and the implications for national attitudes to the Grand Ethiopian Renaissance Dam (GERD). In Grandi's words, "narratives over waters provide the linkage between definitions and practices, meanings and actions, and emerge from configurations of social relations that reflect, and in turn influence, interests and objectives of competing actors at different levels" (p.188). Others, such as Rusca et al. (2019) and Akhter (2022), employ the concept of imaginaries to examine discursive justifications for large water infrastructure projects.

This diversity of how discourse is understood has led to different ways to do discourse analysis in both hydropolitics and IR. Milliken (1999) was the first to point out the lack of a shared understanding of the best ways to study discourse in IR, concluding that the inattention to issues of discourse research puts discourse analysts at a disadvantage within the research community. In her view, the problem is the lack of a common understanding about how to do discourse analysis in International Relations. For Holzscheiter (2014) and Carta (2019), Milliken's claim will always remain valid due to the very nature of analyzing international relations from a discursive perspective: it cuts across several ontological, epistemological, and methodological lines. In hydropolitics, the diversity of discourse analysis approaches is evident in the work of Trottier (1999), Feitelson (2002), Jägerskog (2003), Jacobs (2010), Dore et al. (2012), Kraak (2012), Zawahri and Hensengerth (2012), Mirumachi (2013), Crow-Miller (2015), Ide and Fröhlich (2015), Menga and Mirumachi (2016), Menga (2017), Zinzani and Menga (2017), Hussein (2017, 2018, 2019), Nagheeby and Warner (2018), Williams (2018), Allouche (2019, 2020), Conker and Hussein (2020), and Sayan et al. (2020), among others. While this conceptual and methodological diversity of discourse is not inherently harmful, it can blur the added value of using discourse as an analytical tool to comprehend transboundary water relations.

11.3 The Added Value of Discourse Analysis in Hydropolitics: Empirical Illustrations

In hydropolitical analysis, the hydro-hegemony framework is undoubtedly one of the most popular analytical lenses for studying transboundary water relations. Since its appearance in 2006 (Zeitoun and Warner 2006), the hydro-hegemony framework has evolved and further developed by focusing on the pillars of power (Earle 2010), the varying and co-existing intensities of conflict and cooperation (Zeitoun and Mirumachi 2008), the influence of discursive and ideational acts (Zeitoun et al. 2011), the concepts of counter and virtual hydro-hegemony (Cascão 2008; Sojamo et al. 2012), and the hydro-social analysis of transboundary water arrangements (Zeitoun et al. 2020). Others, such as Menga (2016), have re-theorized the hydro-hegemony framework to put more emphasis on discourses.

Furthermore, an overview of the first ten years of the hydro-hegemony framework (Warner et al. 2017) shows that this string of hydropolitics scholars has privileged discourse analysis. This is because, in the words of the London Water Research Group, "discourse plays a key role in devising 'empty signifiers'" that "are significant not because of their content but due to their effects in bringing actors together" (p.7). Indeed, such emphasis on discourse has a three-fold justification. First, the discourses constructed by the different actors define the problems associated with transboundary water resources, influencing the spectrum of solutions considered. Second, the deployment of these discourses affects the policy-making process, as they provide the framework in which alternatives are evaluated and selected. Third, the construction and deployment of such discourses can reveal the power relations between actors. Or, as Allouche (2020) puts it, "hydropolitical interactions could, in a sense, be conceived as a battleground for ideas and power" (p.287).

Exploring how discourses are constructed and deployed (and by whom) helps comprehend the complexity of hydropolitics. In the following, empirical illustrations from three of the usual suspects for hydropolitical analysis, i.e. the Nile, the Jordan, and the Mekong river basins, demonstrate the added value of bringing a discursive perspective to transboundary waters.

11.3.1 Empirical Illustrations from the Nile River Basin

The Nile river basin, a 3,180,000 km^2 watershed shared between 11 countries in Africa (Burundi, the Democratic Republic of Congo, Egypt, Eritrea, Ethiopia, Kenya, Rwanda, South Sudan, Sudan, Tanzania, and Uganda), is probably one of the most studied hydropolitics case studies in the world. When Waterbury (1979) first coined the term 'hydropolitics', he was examining the utilization of the Nile's transboundary water resources by Egypt and Sudan. Since then, a cascade of scholars has investigated the hydropolitical interactions in the Nile basin (Swain 2002; Cascão 2009; Nicol and Cascão 2011; Obengo 2016; Tayie 2017; Salman 2017; Tawfik 2019; Bekele 2022; Herrer Fernández 2022).

In their hydropolitical analysis of the Nile, some scholars have turned to discourse analysis to understand the interplay between the framing of issues and the resulting hydropolitical interactions. For instance, Menga (2017) uses discourse analysis, based on both official and non-official documents, to provide an account of how the Ethiopian ruling elite managed to frame the Grand Renaissance Ethiopian Dam (GERD) as a foreign policy matter so that issues related to its social and environmental impact nearly disappear from the national political debate. Similarly, Hanna and Allouche (2018) use a discourse analysis approach to examine the Nile water landscape in Egypt as a symbolic and imaginary construction of the nation-state's geographical identity, and they show how the water nationalism discourse in Egypt has influenced transboundary water relations. Also relying on discourse analysis, Hussein et al. (2022) show how both Egypt and Ethiopia deploy different discursive tools (e.g., the securitization of water and non-water issues) to justify, legitimize, and push forward their water imaginaries of the Nile waterscape.

11.3.2 Empirical Illustrations from the Jordan River Basin

The Jordan river basin, an 18,300 km^2 watershed shared between Lebanon, Syria, Israel, Jordan, and Palestine, has been in the focus of hydropolitics scholars for several decades (Wolf 1995; Selby 2003, 2013; Medzini and Wolf 2004; Zeitoun et al. 2013, 2019). The Jordan river basin has been the subject of many analyses, given the centrality of water issues in the Arab–Israeli conflict (including the Israeli-Palestinian conflict), and most recently, the Syrian Civil War.

In the case of the Jordan basin, many scholars have analyzed discourses in their hydropolitical analysis. Trottier (1999), for instance, when examining the 'hydropolitical constellations' in the West Bank and the Gaza Strip, shows that the Palestinian Authority operated between two parallel sanctioned discourses: one targeting local actors and focusing on Israel as a "thief of water rights", the other targeting international actors and adding a dimension of institution building to the discourse on Israel. Similarly, Feitelson (2002) uses discourse analysis to examine the internal Israeli water politics, concluding that shifts in the intra-Israeli water discourse between 1995 and 2000 were not only a shift in the discourse coalitions but also the preparation of a new Israeli water master plan and the win-sets available for the Israeli-Palestinian water negotiations. Hussein (2018) has added another element to the hydropolitical analysis of the Jordan river by showing the impact of water scarcity discourses on transboundary water governance, particularly on the relationships between the Jordanian government and the Syrian, Israeli, and Saudi governments.

11.3.3 Empirical Illustrations from the Mekong River Basin

The Mekong river basin, with a 795,000 km^2 watershed covering China, Myanmar, Thailand, Lao PDR, Cambodia, and Viet Nam, is one of Southeast Asia's most important transboundary rivers. Its hydropolitics has also been analyzed extensively (Li et al. 2011; Biba 2012; Suhardiman et al. 2012; Middleton and Allouche 2016; Mirumachi 2020; Zhang and Li 2020; Tran and Tortajada 2022). Interestingly, in their call to rethink transboundary waters and advance the notion of critical hydropolitics, Sneddon and Fox (2006) drew on the case study of the Mekong river basin.

Subsequently, some scholars have relied on discourse analysis to understand the hydropolitical interactions within the Mekong. Zawahri and Hensengerth (2012) employ the concept of sanctioned discourse to show the tactics of Chinese environmental non-governmental organizations to influence domestic policy for environmental protection. Williams (2018) also applies discourse analysis to the transboundary water dynamics of the Mekong, identifying infrastructure, poverty, development, and climate change as the major narratives, and the regional organizations (i.e., the Mekong River Commission, the Greater Mekong Subregion, and the Lancang-Mekong Cooperation mechanism) as the central fora for institutionalizing such discourses. Most recently, Wang et al. (2022) analyze China's water governmentality in the Lancang-Mekong by studying the official and popular discourses regarding transboundary river governance, uncovering how alternative narratives cast China as a hydro-hegemon that poses threats to downstream riparians.

11.4 A Framework for Systematic Discourse Analysis in Hydropolitics

Despite its added value, discourse analysis remains unexplored, operating as a "black box" in hydropolitical studies, under-elaborated and lacking systematicity and transparency. As Greckhamer and Cilesiz (2014) explain, "although interpretation and judgement [...] are necessary, applying systematic analysis methods and conducting rigorous analysis grounded in epistemological and theoretical assumptions of discourse analysis aid in establishing the trustworthiness of these interpretations and rendering defensible knowledge claims" (p.425). In our recent piece 'Exploring discursive hydropolitics: a conceptual framework and research agenda' (Bréthaut et al. 2022), we put forward a framework for a systematic analysis of discourses in hydropolitics, distinguishing between different categories of discourses (dominant, institutionalized, hegemonic, sanctioned) and discursive processes (structuration, institutionalization, and legitimization). However, there is still one challenge that need to be resolved: operationalizing the conceptual model for real-world, case study applications.

In such operationalization endeavors, we can draw inspiration from Marteen Hajer's argumentative approach to discourse analysis, where hydropolitics becomes an argumentative 'struggle', where actors not only try to make others see the problems according to their views, they also seek to position them in a specific way. Thus, transboundary water conflict is not only a conflict over which sorts of action should be taken but also a conflict over the meaning of physical and social phenomena linked to the transboundary water resources. Analyzing hydropolitical interactions thus entails analyzing the argumentative interactions between actors.

In his book 'The politics of environmental discourse: ecological modernization and the policy process', Hajer (1995) starts by pointing out that discourse analysis consists in "analysing the ways in which certain problems are represented, differences are played out, and social coalitions on specific meanings somehow emerge" (p.44). In this sense, discourse can be seen as the "ensemble of ideas, concepts, and categories through which meaning is given to social and physical phenomena, and which is produced and reproduced through an identifiable set of practices" (Hajer 2006:67). From this definition, three critical points emerge. First, discourse is not restricted to the language we speak or write. It is about the ideas, concepts, and categories that drive how we speak and write. Second, discourse is not only about these ideas, concepts, and categories but also about the social practices embedding them. In other words, it includes the larger system of thought that makes certain things 'thinkable' or 'sayable'. Third, discourse is both produced (a product) and productive (a cause). It (re)produces power relationships and social structures, sometimes reinforcing them, other times challenging them while others are ignored or marginalized.

To operationalize how meaning is conveyed, Hajer proposes the middle-range concept of storylines. In his words, a storyline is a condensed statement summarizing complex narratives, used by people as a shorthand in discussions (Hajer 2005). A storyline thus combines elements from various discourses into a more or less coherent whole and becomes the "medium through which actors try to impose their view of reality on others, suggest certain social positions and practices, and criticize alternative social arrangements" (Hajer 1993:47). In other words, because storylines summarize complex phenomena, they can be used by actors from different backgrounds to fit themselves into the narrative and convey mutual understanding. By taking the storylines as our analytical unit, the modifications in the elements of a discourse can be traced.

Furthermore, to operationalize how actors convene around meaning, Hajer proposes another middle-range concept: discourse coalitions. For him, "in the struggle for discursive hegemony, coalitions are formed among actors [...] that for various reasons are attracted to a specific (set of) storylines" (Hajer 1995:65). Hajer defines discourse coalitions as the ensemble of (1) a set of storylines; (2) the actors who utter these storylines; and (3) the practices in which this discursive activity is based. He emphasizes the "identifiable set of practices" aspect of coalitions, as these practices represent the embedded routines and mutually understood rules and norms that provide coherence to social life (2005:302).

Hajer's approach to discourse analysis provides two concepts, storylines and discourse coalitions, to understand how problems and their solutions are constructed, and how the opposing positions are perceived. Therefore, following Hajer's approach, discourse analysis consists in examining how storylines convey meaning and how discourse coalitions convene around meaning. These two concepts can constitute a solid starting point for developing a systematic, operational discourse analysis framework for hydropolitical research.

Initially conceived for environmental politics, Hajer's approach has also been applied to intra-national hydropolitical analyses. For instance, Whaley and Weatherhead (2014) took inspiration from Hajer, using the concept of interpretative repertoires to analyze the dynamics between farmers and water managers in England and understand the potential for water resources co-management. Their findings reveal that the relative dominance of the competition and conflict repertories in the discourses of farmer groups and water managers indicate a low level of trust among them and a power dynamic favoring individualism and opposition. Kurki et al. (2016) employ Hajer's concept of storylines and discourse coalitions to examine the interactions formed around an artificial groundwater recharge project in Finland. Through four different storylines (i.e. water stress, local rights, environmental changes, and health risk), actors positioned themselves into coalitions to support their views in the mercury debate. Similarly, Tur et al. (2018) also draw upon storylines and discourse coalitions to analyze the debate around hydroelectric dam building in Brazil, taking the case of the Primavera dam in the Parana river basin. One of the most striking results of their analysis is that the four storylines that were identified based on the actors' arguments, using Principal Component Analysis, are not employed by the groups of people directly affected by the building of the dam, revealing that the other groups do not share their heavy symbolic and emotional arguments. Shah and Mathur (2019) rely on Hajer's framework to analyze contestations around water allocation during a 2016 drought event in the state of Maharashtra, India. They argue that the Indian Premier League (IPL) vs drought case, which caused a major controversy around water allocation during an extreme drought event, represents a clash between two competing storylines: water as an economic good and water as a human right.

Furthermore, scholars have recently started to use Hajer's argumentative approach in transboundary hydropolitics. For instance, Isaac and de Loë (2020) rely on storylines and discourse coalitions to analyze water quality policy discourse in the Erie Basin, a transboundary lake between Canada and the US. Their analysis reveals two storylines framing the issue of lake eutrophication (eutrophication due to external factors and eutrophication due to weak governance) and assigns different responsibilities to the actors involved. Sayan et al. (2020) apply Hajer's framework to Lake Chad, a transboundary water system shared by Niger, Nigeria, Chad, and Cameroon. Their analysis of the interbasin water transfer project from the Congo river to Lake Chad shows that non-state actors have been instrumental in shaping transboundary water interactions, mainly through the discursive practices of agenda setting, knowledge construction, securitization, and issue linkage. Ezbakhe et al. (2021) also rely on storylines and discourse coalitions to examine the evolution of narratives surrounding the governance of the Sixaola river basin, which is shared between Costa Rica and Panama. By looking at the dominant storylines from the 1970s to 2015, the analysis shows that international organizations have successfully framed the Sixaola 'problematique' as local while infusing it with a consistent regional Central American perspective.

As seen above, Hajer's framework, with its middle-range concepts of storylines and discourse coalition, can be a good first step in building a systematic approach to discourse analysis in hydropolitics. Taking storylines as the primary analytical variable, it can be examined what is viewed as a relevant problem and trace its emergence. Then, by looking at the actors behind the storylines (i.e. discourse coalitions), the who behind discourse construction, reproduction, and legitimation can be examined, and, most importantly, how they form discourse coalitions at multiple levels and scales. Therefore, to go back to the conceptual framework of Bréthaut et al. (2022) for discursive hydropolitics, understanding the processes of discourse structuration and legitimation requires looking at the link between storylines and actors, in particular, whether many actors use a specific storyline to conceptualize the world and whether power elites endorse a particular storyline. On the other hand, to understand the process of discourse institutionalization, we need to look at the link between storylines and actors, particularly whether a particular storyline manifests in specific institutional arrangements.

What is now missing is a more structured and transparent way to identify and examine Hajer's middle-range concepts of storylines and discourse coalitions. Indeed, while Hajer provides theoretical justifications for his approach to discourse analysis, he does not offer much practical application of the two key concepts (storylines and discourse analysis). While it is clear that there will not be a one-size-fits-all solution to the analysis of discourse, it is crucial to provide more systematicity and transparency when using discourse analysis as a methodology for hydropolitics. This will highlight its added value and replicability in different contexts and case studies.

11.5 Conclusion

Discourse analysis is not an exact science. It not only crosses ontological, epistemological, and methodological lines, it also lacks a unitary theoretical framework behind it. A review of scholarship on hydropolitics reveals that discourses have become a powerful analytical tool, with several approaches followed to analyze discourse. This multiplicity is not an issue in itself, as it shows the flexibility and fluidity of discourse analysis. The problem is the lack of systematicity. Even when Marteen Hajer's argumentative approach to discourse analysis is employed, i.e., undertaking the analysis through the concepts of storylines and discourse coalitions, it is often without explicitly clarifying the methodological steps for identifying such storylines and discourse coalitions. Storylines are generally reported as 'emerging' or 'appearing', which might lead scholars to question the trustworthiness and rigorousness of the results and prevent them from employing Hajer's discourse analysis approach in hydropolitics.

Therefore, there is still room for operationalizing Hajer's discourse analysis and providing a structured, practical application of its storylines and discourse coalition concepts. When applied to hydropolitical studies, such operationalization will allow the examination of multiple case studies guided by a single analytical and operational framework, thus making it possible to uncover patterns in hydropolitical interactions over different transboundary waters. However, instead of a universal recipe for discourse analysis, future research should provide a more transparent, practical approach for using storylines and discourse coalitions as the key analytical variables in hydropolitics.

References

Akhter M (2022) Dams, development, and racialised internal peripheries: hydraulic imaginaries as hegemonic strategy in Pakistan. Antipode 54(5):1429–1450

Allouche J (2019) State building, nation making and post-colonial hydropolitics in India and Israel: visible and hidden forms of violence at multiple scales. Polit Geogr 75:102051

Allouche J (2020) Nationalism, legitimacy and hegemony in transboundary water interactions. Water Altern 13(2):286–301

Bekele TM (2022) The Nile and Eritrea: the nexus between two major conflict issues of Northeast Africa. Ethiop J Soc Sci 8(1)

Biba S (2012) China's continuous dam-building on the Mekong River. Journal of Contemporary Asia 42(4):603–628

Bréthaut C et al (2022) Exploring discursive hydropolitics: a conceptual framework and research agenda. Int J Water Resour Dev 38(3):464–479. https://doi.org/10.1080/07900627.2021.1944845

Carta C (2019) "A rose by any other name": on ways of approaching discourse analysis. Int Stud Rev 21(1):81–106

Cascão AE (2008) Ethiopia–challenges to Egyptian hegemony in the Nile Basin. Water Policy 10(S2):13–28

Cascão AE (2009) Changing power relations in the Nile river basin: Unilateralism vs. cooperation? Water Altern 2(2):245–268

Conker A, Hussein H (2020) Hydropolitics and issue-linkage along the Orontes River Basin: an analysis of the Lebanon-Syria and Syria-Turkey hydropolitical relations. Int Environ Agreements: Polit, Law Econ 20(1):103–121

Crow-Miller B (2015) Discourses of deflection: the politics of framing China's South-North Water Transfer Project. Water Altern 8(2):173–192

Diez T (1999) Speaking 'Europe': the politics of integration discourse. J Eur Publ Policy 6(4):598–613

Dore J et al (2012) A framework for analysing transboundary water governance complexes, illustrated in the Mekong Region. J Hydrol 466–467:23–36

Dryzek JS (1997) The politics of the Earth: environmental discourses. Oxford University Press, Oxford and New York

Earle A (ed) (2010) Transboundary water management: principles and practice (1st ed). Routledge. https://doi.org/10.4324/9781849776585

Ezbakhe F et al (2021) Deconstructing power dynamics and prevailing discourses in hydropolitics: the case of the Sixaola river basin. WIT Trans Ecol Environ 251:1–9

Feitelson E (2002) Implications of shifts in the Israeli water discourse for Israeli-Palestinian water negotiations. Polit Geogr 21(3):293–318

Fierke KM (2002) Links across the abyss: language and logic in international relations. Int Stud Quart 46(3):331–354

Furlong K (2006) Hidden theories, troubled waters: international relations, the "territorial trap", and the Southern African Development Community's transboundary waters. Polit Geogr 25(4):438–458

Geneva Water Hub (2021) What is hydropolitics? Examining the meaning of an evolving field. https://www.genevawaterhub.org/fr/ressource/bibliography-what-hydropolitics-examining-meaning-evolving-field. Accessed 11 Jan 2023

Grandi M (2020) Hydropolitics. In: Oxford Research Encyclopedia of Environmental Science. https://doi.org/10.1093/acrefore/9780199389414.013.644

Grandi M (2021) Images of the Nile: how competing narratives frame water disputes. In: Cortesi L, Joy KJ (eds) Split waters: the idea of water conflicts. Routledge, pp 176–193

Greckhamer T, Cilesiz S (2014) Rigor, transparency, evidence, and representation in discourse analysis: challenges and recommendations. Int J Qual Methods 13(1):422–443

Hajer, Maarten A. (1993) Discourse coalitions and the institutionalization of practice: the case of acid rain in Great Britain. In: Fischer F, Forester J (eds) The argumentative turn in policy analysis and planning. New York, USA: Duke University Press, pp. 43–76. https://doi.org/10.1515/9780822381815-003

Hajer M (1995) The politics of environmental discourse: ecological modernization and the policy process. Oxford University Press, Oxford

Hajer M (2005) Coalitions, practices, and meaning in environmental politics: from acid rain to BSE. In: Howarth DR, Torfing J (eds) Discourse theory in European politics. Palgrave Macmillan, London. https://doi.org/10.1057/9780230523364_13

Hajer M (2006) Doing discourse analysis: coalitions, practices, meaning. In: van der Brink M, Metze T (eds) Words matter in policy and planning—discourse theory and method in the social sciences. Koninklijk Nederlands Aardrijkskundig Genootschap, Utrecht, pp 65–74
Hanna R, Allouche J (2018) Water nationalism in Egypt: state-building, nation-making and Nile hydropolitics. In: Menga F, Swyngedouw E (eds) Water, technology and the nation-state. Routledge, London-New York, pp 81–95
Herrer Fernández S (2022) Water conflict analysis: the Nile River dispute. Revista Análisis Jurídico-Político 4(8):83–112
Holzscheiter A (2014) Between communicative interaction and structures of signification: discourse theory and analysis in International Relations. Int Stud Perspect 15(2):142–162
Hussein H (2016) An analysis of the discourse of water scarcity and hydropolitical dynamics in the case of Jordan. PhD thesis. University of East Anglia (UEA)
Hussein H (2017) Whose 'reality'? Discourses and hydropolitics along the Yarmouk River. Contemporary Levant 2(2):103–115
Hussein H (2018) Lifting the veil: unpacking the discourse of water scarcity in Jordan. Environ Sci Policy 89:385–392
Hussein H (2019) Yarmouk, Jordan, and Disi basins: examining the impact of the discourse of water scarcity in Jordan on transboundary water governance. Mediterranean Politics 24(3):269–289
Hussein H et al (2022) Small is beautiful but not trendy: understanding the allure of big hydraulic works in the Euphrates-Tigris and Nile waterscapes. Mediterranean Politics 27(3):297–320
Ide T, Fröhlich C (2015) Socio-environmental cooperation and conflict? A discursive understanding and its application to the case of Israel and Palestine. Earth System Dynamics 6:659–671
Isaac BN, de Loë E (2020) Eutrophication and water quality policy discourse in the Lake Erie Basin. Water Altern 13(3):800–821
Jacobs I (2010) A fragile hegemon, a fragile hegemonic discourse: a critical engagement with the hydropolitical complex and implications of South Africa's hydropolitical environment for Southern Africa. Afr Secur 3(1):21–45
Jägerskog A (2003) The power of the "sanctioned discourse": a crucial factor in determining water policy. Water Sci Technol 47(6):161–166
Julien F (2012) Hydropolitics is what societies make of it (or why we need a constructivist approach to the geopolitics of water). Int J Sustain Soc 4(1–2):45–71
Kraak E (2012) Diverging discourses on the Syr Darya. Geogr, Environ, Sustain 5(2):36–50
Kurki V et al (2016) Clashing coalitions: a discourse analysis of an artificial groundwater recharge project in Finland. Local Environ 21(11):1317–1331
Li Z et al (2011) Regional hydropolitics of the transboundary impacts of the Lancang cascade dams. Water Int 36(3):328–339
Medzini A, Wolf AT (2004) Towards a Middle East at peace: hidden issues in Arab-Israeli hydropolitics. Int J Water Resour Dev 20(2):193–204
Menga F (2016) Reconceptualizing hegemony: the circle of hydro-hegemony. Water Policy 18(2):401–418. https://iwaponline.com/wp/article-abstract/18/2/401/20486/Reconceptualizing-hegemony-the-circle-of-hydro?redirectedFrom=fulltext
Menga F, Mirumachi N (2016) Fostering Tajik hydraulic development: examining the role of soft power in the case of the Rogun Dam. Water Altern 9(2):373–388
Menga F (2017) Hydropolis: reinterpreting the polis in water politics. Polit Geogr 60(Sept):100–109
Middleton C, Allouche J (2016) Watershed or powershed? Critical hydropolitics, China and the 'Lancang-Mekong cooperation framework.' Int Spect 51(3):100–117
Milliken J (1999) The study of discourse in International Relations: a critique of research and methods. Eur J Int Rel 5(2):225–254
Mirumachi N (2013) Securitising shared waters: an analysis of the hydropolitical context of the Tanakpur Barrage project between Nepal and India. Geogr J 179(4):309–319
Mirumachi N (2020) Informal water diplomacy and power: a case of seeking water security in the Mekong River basin. Environ Sci Policy 114(3):86–95

Müller M (2008) Reconsidering the concept of discourse for the field of Critical Geopolitics: towards discourse as language *and* practice. Polit Geogr 27(3):322–338

Nagheeby M, Warner J (2018) The geopolitical overlay of the hydropolitics of the Harirud River Basin. Int Environ Agreements: Polit, Law Econ 18:839–860

Nicol A, Cascão AE (2011) Against the flow: new power dynamics and upstream mobilisation in the Nile Basin. Rev Afr Polit Econ 38(128):317–325

Obengo JO (2016) Hydropolitics of the Nile: the case of Ethiopia and Egypt. Afr Secur Rev 25(1):95–103

Rusca, M. et al. (2019). Space, state-building and the hydraulic mission: crafting the Mozambican state. Environment and Planning C: Politics and Space, 37(5):868–888

Saklani U et al (2020) Hydro-energy cooperation in South Asia: prospects for transboundary energy and water security. Environ Sci Policy 114:22–34

Salman SM (2017) The Nile basin cooperative framework agreement: disentangling the Gordian knot. In: Yihdego Z, Rieu-Clarke A, Cascão AE (eds) The Grand Ethiopian Renaissance Dam and the Nile Basin. Routledge, London and New York, pp 18–40

Sayan RC et al (2020) Soft power, discourse coalitions, and the proposed interbasin water transfer between Lake Chad and the Congo River. Water Altern 13(3):752–778

Selby J (2003) Dressing up domination as 'cooperation': the case of Israeli-Palestinian water relations. Rev Int Stud 29(1):121–138

Selby J (2013) Cooperation, domination and colonisation: the Israeli-Palestinian Joint Water Committee. Water Altern 6(1):1–24

Shah A, Mathur N (2019) Contestations around water allocation during a climate crisis in India: the case of 'IPL vs. drought'. Glob Environ Chang 57:101927

Sneddon C, Fox C (2006) Rethinking transboundary waters: a critical hydropolitics of the Mekong basin. Polit Geogr 25(2):181–202

Sojamo S et al (2012) Virtual water hegemony: the role of agribusiness in global water governance. Water Int 37(2):169–182

Suhardiman D et al (2012) Scalar disconnect: the logic of transboundary water governance in the Mekong. Soc Nat Resour 25(6):572–586

Swain A (2002) The Nile River Basin Initiative. SAIS Rev 22(2):293–308

Tran TA, Tortajada C (2022) Responding to transboundary water challenges in the Vietnamese Mekong Delta: in search of institutional fit. Environ Policy Gov 32(4):331–347

Tayie MS (2017) The hydropolitics of the Nile River basin. In: Negm AM (ed) The Nile River. Springer, Cham, pp 599–654

Tawfik R (2019) Beyond the river: elite perceptions and regional cooperation in the Eastern Nile Basin. Water Altern 12(2):655–675

Trottier J (1999) Hydropolitics in the West Bank and Gaza Strip. PASSIA—Palestinian Academic Society for the Study of International Affairs, Jerusalem

Tur AA et al (2018) Discourse analysis of the debate on hydroelectric dam building in Brazil. Water Altern 11(1):125–141

Wang RY et al (2022) China's water governmentality and the shaping of hydrosocial territories in the Lancang-Mekong Region. China Q 252:1233–1255

Warner J et al (2017) Transboundary 'hydro-hegemony': 10 years later. Wiley Interdiscip Rev Water 4(6):e1242

Waterbury J (1979) Hydropolitics of the Nile Valley. Syracuse University Press, Syracuse

Whaley L, Weatherhead EK (2014) Competition, conflict, and compromise: three discourses used by irrigators in England and their implications for the co-management of water resources. Water Altern 8(1):800–819

Williams JM (2018) Stagnant rivers: transboundary water security in South and Southeast Asia. Water 10(12):1819

Wolf AT (1995) Hydropolitics along the Jordan River: scarce water and its impact on the Arab-Israeli conflict. United Nations University Press, Tokyo

Zhang H, Li M (2020) China's water diplomacy in the Mekong: a paradigm shift and the role of Yunnan provincial government. Water Int 45(4):347–364

Zawahri NA, Hensengerth O (2012) Domestic environmental activists and the governance of the Ganges and Mekong Rivers in India and China. Int Environ Agreements: Polit, Law Econ 12(3):269–298

Zeitoun M, Warner J (2006) Hydro-hegemony: a framework for analysis of trans-boundary water conflicts. Water Policy 8(5):435–460

Zeitoun M, Mirumachi N (2008) Transboundary water interaction I: reconsidering conflict and cooperation. Int Environ Agreements: Polit, Law Econ 8(4):297–316

Zeitoun M et al (2011) Transboundary water interaction II: the influence of 'soft' power. Int Environ Agreements: Polit, Law Econ 11(2):159–178

Zeitoun M et al (2013) Hydro-hegemony in the Upper Jordan waterscape: control and use of the flows. Water Altern 6(1):86–106

Zeitoun M et al (2019) The Yarmouk tributary to the Jordan River I: agreements impeding equitable transboundary water arrangements. Water Altern 12(3):1064–1094

Zeitoun M et al (2020) Analysis for water conflict transformation. Water Int 45(4):365–384

Zinzani A, Menga F (2017) The circle of hydro-hegemony between riparian states, development policies and borderlands: evidence from the Talas waterscape (Kyrgyzstan-Kazakhstan). Geoforum 85:112–121

Chapter 12
The Upstream–Downstream Structural Disparity and Transboundary Challenges of Public Health

Péter Marton

Abstract This chapter provides an overview of the public health challenges of international concern that are connected to the interaction between human activity, natural processes, and the broader environment, as it is mediated by rivers. The chapter is based on the premise that a One Water Basin/One Health perspective is necessary for a comprehensive appraisal of the subject, and a discussion of the latter is presented to lay the grounds for the ensuing analysis. The causal mechanisms of transboundary impact are then identified from the available empirical literature, with the aim of constructing analytical categories for assessment. The resulting categories are considered in order to be able to judge how the public health dimension of upstream–downstream interactions relates to the upstream–downstream structural disparity that fundamentally shapes relations between riparian states, and how it consequently adds to the incentive structure of international cooperation in water basin governance.

Keywords Hydropolitics · Public health · River basins · Riverine waters · Upstream–downstream structural disparity · Water-mediated diseases

12.1 Introduction: One Health in Water Basins

The idea of One Health is based on the understanding that when working towards the general goals of public health, it is indispensable to address the interconnectedness of the health of humans and (non-human) animals, even if the health of the human population is prioritized (Mackenzie and Jeggo 2019). Even plants may fall under the scope of this concept, not only as food sources for animals and humans but also, for example, due to their role in shaping the climate. Once the implications of all possible interactions are taken into account, the realization must follow that a One Health approach needs to consider the integrity of the entire ecosystem. Certain

P. Marton (✉)
Corvinus University of Budapest, Budapest, Hungary
e-mail: peter.marton@uni-corvinus.hu

K. Szálkai and M. Szalai (eds.), *Theorizing Transboundary Waters in International Relations*, Springer Water, https://doi.org/10.1007/978-3-031-43376-4_12

practical contradictions between the immediate needs of humans and the long-term interest of the environment can only be resolved and/or transcended through such an ecosystem-centric perspective. For instance, some animal populations, such as mosquitos, need to be controlled for the sake of (human) public health, but they need to be controlled in a way that avoids or at least minimizes the environmental damage and ecosystem disruption resulting from it.

The origins of the One Health concept date back to the work of nineteenth-century German physician Rudolf Virchow, whose conceptualization of "zoonosis" (the emergence of disease upon interspecies transmission from a non-human animal to a human) gave rise to the doctrine of One Medicine (Gyles 2016:345). This spurred interest in the comparative pathology of animals and humans, as well as in establishing a veterinary branch of public health. Another tributary idea of the One Health framework is the basic notion of "world health" as a global public good, indivisible and of benefit to all—with reference to the interconnectedness, and the consequent interdependence, of the health of all human populations worldwide, basically as "public health" writ large as a "matter of scale" (Koch 2014:1014).

It is common to point to this interdependence as the reason for a need to coordinate health policies globally, and to share at least some of the burdens of striving for better health outcomes worldwide. When examined in the context of particular medical conditions, however, this interdependence is only present clearly and directly in the case of infectious diseases. In carrying human hosts, highly transmissive infectious disease agents can easily leap from one continent to another in the age of jet travel. Even weakly transmissive infectious disease agents may affect people beyond the traditionally endemic areas, aided by processes of human migration and climate change, or if travelers returning from endemic areas import infection to non-endemic regions.

At the same time, the interdependence of world health is at best indirect and may only exist through complex mechanisms in other contexts. For instance, the large chronic disease burden of societies in high-income countries is only extra-medically linked to the economic, social, and public health conditions in other parts of the world. While it is absolutely warranted, or even necessary, to consider, related to the comparative prevalence of chronic diseases, the historical determinant of health in how colonialism contributed to a world of vast health disparities today, where the disease burden of different societies can vary to such a great extent, a more narrow definition of health interdependence may better suit the purposes of the present inquiry.

As this chapter focuses on the public health challenges of international concern that arise in the context of hydro-political systems, interdependence here largely refers to the direct health impact of the effects transmitted by the relatively closely interacting structures of terrain, flow of water, human and animal populations (and their pathogens), and human activity (including political agency), confined within the territory and related geographical barriers of water basins the territory of which is divided between two or more countries.

Hydropolitics, arising from the complex of these structures, may be defined as revolving around "a set of issues and mechanisms with important, and, at the same time, peculiar implications for the study of security." There usually arise in these systems of interaction various insecurities connected to the "vertical, relational positioning of 'upstream' and 'downstream' actors within water basins" and the actual "mechanisms by which riverine water-related [interactions] may generate insecurities for different territorial units" (Marton and Szálkai 2017:4–5). These riverine water-related interactions do not only have an impact from upstream to downstream areas, although the power of gravity certainly channels much of the impact that way. In Moellenkamp's assessment (2007:1409), there is "a unidirectional externality" in the upstream–downstream relationship (elsewhere referred to by the same author as "asymmetrical") because "the upstream country profits from the water resource whereas the downstream country suffers from pollution and … restricted water use." Yousef refers to this set of problems as the "river-sharing game," where "Naturally, D [downstream country] has no control over the size of [water inflow to its territory] due to the river's geographic configuration but U [upstream country] can control it in several ways" (Yousef 2021:7).

This basic structural imbalance is in some cases offset by the fact that many (although by far not all) upstream countries are landlocked, and may hence be vulnerable in several ways, incentivizing them to seek cooperation with their neighbors. Further complicating the seemingly straightforward implications of the upstream–downstream structural disparity, the downstream countries that have access to the sea may benefit from saltwater desalination to mitigate their water scarcity (Aviram et al. 2014).

The cumulative significance of the upstream–downstream effects for public health, as well as the need to understand their role in causing and preventing some of the burden of water-mediated diseases, merits attention in a One Water Basin/One Health framework. In turn, the overall balance of these effects is of interest with a view as to how much upstream countries may be incentivized by them, if at all, to cooperate in addressing broader downstream concerns due to their exposure to the various effects transmitted along rivers. The natural expectation must be that upstream and downstream areas are asymmetrically exposed to such effects. For any impact to emerge upstream, some vehicle of the effect in question must defy gravity. If there is indeed an absence of major upstream exposure, it would confirm that equitable water basin governance can only be sufficiently incentivized by extra-sectoral factors, i.e. factors not connected (or only indirectly connected) to the direct benefits of cooperation in water governance. For instance, in the field of trade, alliance politics, or by way of compellence of some kind.

12.2 Transboundary Public Health Effects Mediated by Rivers: Mechanisms

Following in the footsteps of Szálkai (2022:2–3), this chapter uses a broad definition of water-mediated disease. It focuses on water-borne, water-washed, water-related, and water-based diseases caused by microbiological agents, as well as water-distributed diseases that emerge as a result of riverine water pollution.[1] Keeping with the One Health approach discussed above, river-mediated effects are also considered with a view to the health of some animal populations. Relevant examples of these effects are identified from the empirical literature.

Given that the primary aim of this chapter is to offer a conceptual discussion rather than empirical findings as such, it is important to emphasize that interactions between rivers, human and animal populations, pathogens, and climatic and other environmental factors are very complex, and only bear limited generalization across water basins. No observation of the balance of upstream, downstream, and other effects within a particular basin should necessarily be valid beyond the surrounding watershed.

This chapter discusses eight mechanisms of interest to illustrate the significance of transboundary public health impacts.

- *Deprivation.* Water abstraction causing scarcity downstream.
- *Spoiling.* Water abstraction causing water levels to drop downstream, resulting in the proliferation of pathogens under some circumstances.
- *Vector amplification.* Holding water upstream, resulting in the proliferation of vectors of disease and the consequent increase in the prevalence of mosquito-borne and other diseases in the affected area.
- *Chronic pollution.* Upstream activities, including protracted contamination, affecting the quality of water downstream.
- *Acute pollution.* Major incidents of pollution impacting areas downstream.
- *Disease exportation/importation (concerning fish).* Epidemics of fish.
- *Flood amplification.* Catastrophic floods destabilizing ecosystems, leading to zoonotic incidents and epidemics, with intense floods made more likely by climatic factors and human activity throughout a water basin.
- *Policy coordination failure.* Lack of sufficient cooperation in disease control in cross-border endemic regions.

As pointed out above, much of the impact is channeled from upstream towards downstream areas, but there are exceptions, too. Holding water upstream in large

[1] To clarify the terminology, water-borne diseases are caused by pathogens present in surface waters, ingested by suitable hosts; water-washed diseases are caused by the lack of water in adequate quantities for hygiene purposes, which could make such diseases preventable; water-based diseases are caused by agents (e.g. parasites such as the Guinea worm) that live a part of their lifecycle in surface waters; water-related diseases are mediated by vectors (e.g. mosquitos) that rely on the presence of water to reproduce; while water-distributed diseases are caused by contaminants in the water.

water reservoirs accounts for local effects, some of which may by exerted in a transboundary area. Epidemics of fish can spread wherever fish populations swim, or where fish breeders or anglers contribute to their exportation/importation. Lack of coordination and insufficient effort in terms of disease control may be present in upstream as well as downstream areas. Finally, catastrophic flooding may arise in part due to human activity, including such activity increasing vulnerability downstream. These issues are further detailed below.

Water abstraction causing scarcity downstream. Scarcity, or the shortage of drinking water and water available for other uses, is a general obstacle to life and economic development in the affected regions. Agricultural produce suffers as a result, and people's livelihood is endangered. Water scarcity can thus result in a more general crisis of critical supplies, including food insecurity. Considering the more direct effects, water shortages do not simply lead to an increased prevalence of water-washed diseases due to poorer hygiene, as they inevitably induce the consumption of more unsafe water and consequently increase the prevalence of water-borne diseases. In areas protractedly exposed to these conditions, more diarrheal disease and malnutrition can translate into the stunted physical and cognitive development of children, beyond the immediate morbidity and mortality burden (Grantham-McGregor and Fernald 1997; Berkman et al. 2002). Meanwhile, time spent collecting water also increases, affecting the schooling of children who may be sent on water-related errands. Furthermore, the inadequate availability of water at schools can also lead to female students dropping out during menses (Tarrass and Benjelloun 2012:242). A prime example of a country exposed to the loss of access to water due to upstream dam construction is Iraq, which has in recent years seen a shrinking of cultivated land, increasing drinking water shortage and desertification as a result.

Water abstraction causing water levels to drop downstream. Jutla et al. (2013:598) built a hydro-climatological model to study the occurrence of epidemic cholera, in which elevated water temperatures contribute to bacterial growth, including the growth of the pathogen *Vibrio cholerae*, followed by above normal rainfall, e.g., the monsoon rains in South Asia. These rains lead to the overflowing of sewers and the destruction of sanitation infrastructure. Many aspects of this model have been empirically verified, including the role of elevated temperatures in the months preceding major outbreaks of cholera. On the other hand, this is not an entirely natural process, since it requires poor sanitation infrastructure and a population relying on surface waters for its daily consumption needs. Decreased water levels also play a role (Jutla et al. 2013:601), allowing for the increased warming and evaporation of the remaining shallower body of river water, along with its salination, further favoring bacterial growth. Even if the model by Jutla et al. does not necessarily apply in all epidemic regions of cholera equally, the posited mechanisms are grounds for concern everywhere, including the Ganges Delta region, which stands to be strongly affected by upstream interventions, including prospective dam construction by China to hold the waters of the River Brahmaputra. In addition, given the relationship between water temperature and bacterial growth, a host of other pathogens may also be promoted under these circumstances.

Holding water upstream, resulting in the proliferation of vectors of disease. Dam construction and the related creation of large water reservoirs has long been connected to an increase in the malaria burden near reservoir shores. A recent study by Kibret et al. (2019) confirms that there is an impact, but it also shows that the relationship is not entirely straightforward. There is an absolute growth in the incidence of mosquito-borne diseases, but it is linked to the growth of the population around reservoirs against a backdrop of decreased (population-proportionate) prevalence. Moreover, the overall impact is always mediated by a host of factors. For instance, the impact may decrease with distance from the shoreline, but this may be counteracted by dominant wind patterns conducive to the mosquitos' presence in a particular wind channel. The steepness of the shoreline slope is also found to be a key variable, with steeper slopes presenting less favorable conditions for the breeding of mosquitos. The type of human activity around the reservoir is not indifferent, either: increased use of irrigation for agricultural purposes can worsen the challenges of mosquito control. Meanwhile, mosquito-borne diseases are not the only concern. Studies have shown important adverse health impacts from damming rivers in tropical areas due to the increased burden of schistosomiasis in nearby communities. That is because the snails that are the intermediate hosts of the *Schistosoma* parasite can proliferate unexposed to predation by *Macrobrachium* prawns, their key predators, which are obstructed in their movement (Sokolow et al. 2017). The resulting public health impact may be transboundary—and thus shared internationally—around water reservoirs that stretch across borders between two or more countries, for example, in the case of Lake Nasser, where a larger part of the lake is in Egypt, and a small part is in Sudan.

Upstream activities affecting the quality of water downstream. Upstream activities of just about any kind are sources of persistent contamination in waters downstream, be these related to agricultural, mining, or industrial activities, or merely the production of waste and sewage in the course of everyday human life. Pollution may reach waters as run-off (carried by water from overland), it does not need to be intentionally released directly into a surface water stream. Concerns related to this extend beyond the length of a river, from its source to its confluence with other rivers, or its estuary, delta, or mouth to the sea. As the European Union's widely referenced Water Framework Directive rightly notes, "An effective and coherent water policy must take account of the vulnerability of aquatic ecosystems … near the coast and estuaries or in gulfs or relatively closed seas, as their equilibrium is strongly influenced by the quality of inland waters flowing into them" (WFD 2000:L 327/2). Even the health of fish populations and the livelihood of coastal human and other populations is thus dependent in part on what happens to riverine ecosystems. One of the most important negative health impacts associated with "nonpoint" agricultural runoffs is due to nitrogen and phosphorous reaching major surface waters and creating oxygen-starved zones in them. Increasingly, attention is also paid to the so-called emerging contaminants, all kinds of substances from antibiotics to hormones (for example, related to Concentrated Animal Farming Operations). These are usually present in lower, difficult-to-detect quantities that do, nonetheless, have an impact on biological life, e.g., on the reproductive health of the populations encountered and the growth of antimicrobial resistance in water-borne pathogens (Damewood 2013). One of the latest-recognized sources of concern is the group of carbon fluorides known as "forever chemicals," due to their persistent presence in the environment; these are called PFAS (per- and polyfluoroalkyl substances). These substances have been linked to adverse pregnancy outcomes for both mother and offspring, as well as cancers, such as kidney and testicular cancer, in those exposed to larger doses (Blake and Fenton 2020; Steenland and Winquist 2021).

Major incidents of pollution impacting areas downstream. Mining operations, along with chemical industrial and nuclear power production, are just some of the activities that may be concerning from a downstream perspective and have historically led to major catastrophic (sudden-onset, extraordinary) incidents of pollution. In particular, tailings dam/waste reservoir failures stand out as having frequently resulted in such events. In fact, these are likely to continue to happen in the future, given the accelerating exploitation of the mineral resources of the planet. Examples of such dam failures include the 2000 Baia Mare cyanide spill in Romania, with a devastating impact on the fish populace in the River Tisza, or the Mariana dam disaster in Bento Rodrigues, Brazil, in November 2015, contaminating both Rio Doce and the waters of the Southern Atlantic Ocean. A similar industrial disaster of note was the toxic red sludge spill that happened in Hungary in October 2010 (although with much better contained impact, with the effects downstream limited in range and scope).

Epidemics of fish. Epidemics among fish can be devastating both in terms of population morbidity and mortality and in terms of the economic damage to aquaculture or fish farming. For instance, a single outbreak of Viral Haemorrhagic Septicaemia Virus beginning in 1998 caused the Western European aquaculture industry losses amounting to USD 60 million per year for the duration of the epidemic (Jonkers et al. 2010). One of the currently most concerning infections of carp and other key fish stocks worldwide, the Koi Herpes Virus, emerged around the same time (Pokorova et al. 2005; Bergmann et al. 2020). Some fish pathogens can be transmitted very easily, some stay infectious in the water for hours outside a host, and many can even be transmitted by anglers using the same tools across different waters. Movements of fish are significant here with a view to their downstream-upstream mobility. There is a catch, however: a study of parasite abundance by Blasco-Costa et al. (2013) found a longitudinal gradient, i.e., a decrease of abundance along the downstream-to-upstream continuum. Various factors can be surmised to play a role in this, including the unidirectional flow of the river and host density. For some of the same reasons, viral abundance may be higher downstream, too, even as certain host behavior patterns (or the conditions, such as water temperature, to which pathogens are differentially adapted) may complicate this relationship.[2]

[2] Epidemics of other animals that share the habitat provided by riverine areas (i.e. rivers and their vicinity) could also be discussed here, but this chapter does not do so for the following two reasons: (1) other river-dwelling species are comparatively less significant from a (human) economic perspective; (2) the life and migration of non-river-dwelling species is not channelled by rivers as much as that of fish. Thus, for example, beavers, although they may give rise to considerable economic damage, are ignored here due to the first consideration in that epidemics controlling their population level may be beneficial in some areas, but not without spill-overs, for instance, in the case of

Catastrophic floods destabilizing ecosystems. The aftermath of natural disasters poses an increased risk of disease outbreaks. This is so for various reasons, among which weak or insufficient institutional response, especially in the immediate wake of disasters, as well as displacement-induced population movements may play a prominent role. Flooding and outbreaks of diarrheal disease often occur together, for example due to the contamination of drinking water supplies and the inadequate sanitation infrastructure available in makeshift displacement camps (Watson et al. 2007:2). A connection with mosquito-borne diseases may exist through the increased use of water containers (e.g., rudimentary rainwater collectors), creating more breeding sites for mosquitos in the wake of a disruption or halt to water supplies (Watson et al. 2007:3). Xu et al. (2014) have shown that historically a connection existed between floods and the plague, offering evidence of this regarding past epidemics in China, where the movements of rodents played an important role in the spread of the disease. As this shows, from the One Health perspective, flood-related displacement does not only concern human populations. Overall, given that the frequency and intensity of flooding is influenced by processes upstream (e.g., deforestation), the public health impact of flooding may in some cases be indirectly connected to such factors. Counterintuitively, even the building of dams upstream, while it generally offers benefits in terms of flood control, may, in the case of larger floods, contribute to more intense flooding along the channel downstream of a dam (Ma et al. 2022).

Lack of sufficient coordination in disease control in cross-border endemic regions. The lack of coordination across jurisdictions facing common problems can lead to all kinds of spill-over effects and suboptimal results, and not only in the field of public health. As regards public health, key consequences may include persistent disease prevalence and even the growth of antimicrobial resistance due to uncoordinated treatment in a context characterized by the large-scale cross-border movement of people. This applies to malaria as well as other diseases. Recognition of this challenge has led to the cooperation that was gradually achieved in the Mano River Basin between Sierra Leone, Liberia, and Guinea in the framework of the Mano River Union (Gustavsen et al. 2016). This organization also has some very general objectives, but it has been addressing the public health challenges related to Neglected Tropical Diseases (NTDs) as a priority. This includes the fight against oncocerchiasis or parasitic "river blindness," the spread of which was greatly helped in the past by insufficient collaboration in riverine and wetland border areas. Coordination can make a major difference, for example in the effectiveness of Mass Drug Administration (MDA) campaigns, which aim to cut the parasite lifecycle by targeting entire populations in which individuals may harbor fledgling infections. There are many border regions in Africa where cross-border cooperation is, or would be, similarly necessary. Moreover, achieving sufficient coordination may be a challenge on the sub-state level as well. As Bush et al. put it, "internal borders within a country … can also present considerable programmatic challenges" (2018:i55).

increasing *Giardia* transmission, which can also lead to human cases of Giardiasis. Avian epidemics are ignored due to the second consideration.

12.3 Impact on Upstream–Downstream Disparity

Only three of the mechanisms discussed above function as possible avenues of a negative public health impact originating from downstream areas for upstream areas, namely, vector amplification in border regions with water reservoirs, insufficient vector and disease control in endemic border regions, and epidemics among fish, although, in the case of the latter, this is mitigated by various factors, including the unidirectional flow of water. Overall, downstream-upstream directionality of impact is conceivable mostly due to vector or host mobility, i.e., the movement of snails, mosquitos, fish, and other relevant populations.

Table 12.1 shows the various effects described above that are associated with directionality and labelled as either "on-site," "off-site", or "downstream from site," depending on the territorial scope of their impact. The emerging picture underlines the fundamental asymmetry that favors upstream areas and reinforces upstream–downstream structural disparity.

Beyond those discussed in the previous section, some additional mechanisms may be more tentatively proposed for examination. For instance, groundwater bodies, which may be connected to rivers and other surface waters in complex ways, have their own natural flow, as well as recharge and discharge areas, implying a need for integrated transboundary groundwater management for a truly wholistic approach to water governance (Golovina et al. 2021:2, 11). The contamination of aquifers and other groundwater bodies may travel in directions contrary to the flow of surface waters, turning the upstream–downstream relationship on its head. In fact, such movements of water may even interconnect different water basins, as in the case of the Danube Sinkhole (*Donauversickerung*) that connects the Danube and the Rhein, with the water of the Danube leaking away through caverns and porous karst layers.

Table 12.1 Transboundary public health effects mediated by rivers

Effect	Unidirectionality or bidirectionality	Territorial scope of impact
Deprivation	Unidirectional	Off-site
Spoiling	Unidirectional	Off-site
Vector amplification	n.a.	On-site
Chronic pollution	Unidirectional	Downstream from site
Acute pollution	Unidirectional	Downstream from site
Disease exportation/importation (concerning fish)	Asymmetrically bidirectional (likely skewed towards downstream)	Upstream and downstream from site (likely skewed towards downstream)
Policy coordination failure	Bidirectional	Upstream or downstream from site (originating in the country with a below-par effort)
Flood amplification	Unidirectional	Downstream from site

Furthermore, while the discussion above mostly concerned effects between upstream country “A” and downstream country “B” (in either direction), more diffuse linkages are also conceivable. Both the source and the impacted area may be diffuse, distributed over a larger area. For example, climatic factors, shaped by a complex of global, regional, and local mechanisms, including anthropogenically induced change, impact on precipitation and the uneven availability of water in water basins, amplifying the upstream–downstream structural conflict. Meanwhile, the health of river estuaries and other coastal or inland wetland areas is crucial for long-distance migratory birds (Weidensaul 2021:35–36), whose population health, in turn, is not indifferent to the threat posed by avian influenza strains, pathogens that can spill over into other populations, including the human populace, to become a global threat. It is highly questionable practice in this regard that some metropolises rely on wildlife harassment in key reservoir areas, including the harassment of birds, for “watershed protection,” i.e., the protection of their drinking water supplies from pathogenic contamination, e.g., *E. coli* bacterial contamination (Morris 2007:275). That this seems cost-efficient is in ignorance of the externalities and the implications of the One Health perspective.

These additional mechanisms do not change the fact that the upstream–downstream structural disparity is enhanced rather than mitigated by the public health impacts traveling along surface and other waters within a water basin. The related externalities are not unidirectional but asymmetrical, but this is more than enough to give rise to an unfavorable opportunity/constraint structure for international cooperation and transboundary river basin governance, requiring extra-sectoral compensation for cooperation to be possible and equitable.

12.4 Case Study: Transboundary Impact in the Danube River Basin

The Danube is the second longest river of Europe. Its drainage basin covers an area larger than 800,000 km^2, incorporating all or part of the territory of ten riparian countries (Germany, Austria, Slovakia, Hungary, Croatia, Serbia, Romania, Bulgaria, Ukraine, and Moldova) and some smaller or greater parts of the territory of ten additional countries via its tributaries (Switzerland, the Czech Republic, Slovenia, Bosnia and Herzegovina, Montenegro, Albania, Kosovo, North Macedonia, Italy, and Poland). Managing and mediating demand for the Danube’s water for the purposes of consumption, industrial or agricultural use, fishing, energy production, or navigation, along with interest in the preservation of nature and the protection of the environment, is a cross-border challenge that requires international coordination among the countries concerned and the transnational mobilization of the societies that are affected. Various international frameworks exist for this purpose, including the Danube Commission, founded in 1948, to safeguard appropriate conditions for navigation along the Danube as a major international waterway, or the International

Commission for the Protection of the Danube River (ICDPR), established in 1994 on the basis of the Danube River Protection Convention of 1994.

A TransNational Monitoring Network (TNMN) has also been operating under the ICDPR since 1996 to monitor long-term trends of water quality and signs of pollution across the basin, relying on data collected from over 100 monitoring sites (as of 2016) along the Danube and its major tributaries. While improvements have been recorded during the existence of the TNMN, the lower parts of the river remain in a generally worse condition, conforming to what can be universally expected based on the upstream–downstream structural disparity. It is the downstream areas where the highest concentrations of biodegradable organic matter or toxic metallic contaminants such as cadmium can be found, along with the lowest values of dissolved oxygen (DO).[3] The TNMN Yearbook of 2016 also notes that the improvement of "a decreasing tendency of ortho-phosphate-P concentrations is mostly seen in the upper part of the Danube" (TNMN 2016:16–17). Having said that, a distinction of the upper and the lower parts of the Danube is insufficient for a more nuanced understanding of spatial distributions. For example, cadmium, long-term exposure to which can lead to "damage to the kidneys, liver, testes and prostate … [and] may also cause anaemia, high blood pressure, circulation problems, decalcification of bones and muscle atrophy" (Rzętała 2016:2458), may, according to the findings of a recent study, pose a health risk to children (and to some extent also other residents) who consume peas or radish even in the area of Dunaújváros (Kovács-Bokor et al. 2021:2317), in what may be best described as a mid-stream country, i.e. Hungary.[4]

In a slight deviation from the upstream–downstream dynamic, epidemics of fish in the Danube basin show exactly the kind of complexity that was discussed above. For example, the above-mentioned Koi Herpes Virus (KHV) has been observed to spread to new waters as a result of human activity, e.g. in Croatia (Zrnčić et al. 2020:673), where "[a]nglers are in control of most of the artificial lakes [in the northern part of the country]" and "repopulate artificial and natural lakes as well as open waters with common carp from aquaculture facilities [belonging to the Danube watershed]." Meanwhile, over the previous two decades, a fast-spreading population of the round goby is increasingly seen as threatening to become a key intermediary vector species of transmission for KHV, due to its ability to prevail in both sweet and saltwater environments, and its rapid invasion of habitats along Europe's coasts and rivers, including the Danube basin, where it has migrated from the Black Sea and Sea of Azov region. Human activity has played a role in this as well, e.g. connected to ships transporting round goby between locations with ballast water,[5] but the species itself also shows remarkable versatility and resilience, and it moves, as noted, from downstream to upstream areas wherever it appears (Jin et al. 2020; Kvach et al. 2017).

[3] Lower levels of DO may be due to multiple factors, from higher water temperatures to a higher level of demand for biological oxygen in the presence of untreated sewage in the water.

[4] The Danube drainage basin can be conveniently divided into three sub-basins, as the river passes through major mountain chains at the Devin and Iron Gates, with present-day Hungary lying south of the former and north of the latter.

[5] The round goby can similarly accidentally penetrate fish ponds during the filling process.

As discussed above, climate change amplifies the structural conflict of water scarcity, which is also observable in the Danube basin. The historic droughts of recent years have often caused lakes along lower parts of the Danube to dry up. Again, "lower parts" can also imply mid-stream areas. While global media outlets have taken note of conditions in Romania, mostly related to the spectacle of sunken World War II ships having been revealed by low water levels in the summer of 2022 (Grulovic 2022), the problem manifests in multiple other risks to humans as well as the ecosystem at large. For example, there have been anecdotal reports of the increasing prevalence of avian botulism in drought-stricken eastern Hungary over the same period (Barna 2022), due to the proliferation of *Clostridium botulinum* bacteria in warmer-than-usual receding lakes.

The prospects of continued warming and further droughts in future years has prompted a discussion of how adequate water management ought to imply more of an interest in holding up and storing water flowing through the territory of the riparian countries, including Hungary (Bozzay 2022). Coupled with the potential rise of interest in hydropower production in the region connected to the challenges of the energy transition and the disruptions caused by Russia's war against Ukraine, a conservative assessment must anticipate exacerbated structural conflict over water use. For the time being, however, this is mitigated by the close political, economic, and security integration of the countries of the region, and the seeking of common solutions is strongly incentivized.

12.5 Conclusion

In an overview of intra-basin transboundary challenges of public health, this chapter has demonstrated that the public health dimension of interactions does not attenuate the upstream–downstream structural disparity. This may be considered further evidence that the latter can only be mitigated by factors external to riverine interstate interactions. Various incentives are conceivable but not necessarily present: downstream countries fortunate to have such instruments at their disposal may offer trade advantages, mutually beneficial prospects of economic integration, or other elements of value to an upstream partner. Alternatively, they may compensate for their vulnerability by military strength or political leverage by other means.

There is but a single exception to this basic structural set-up of upstream and downstream countries, namely when the territory of a country holds upstream as well as downstream sections of the constituent surface water streams of a transboundary water basin, with another state holding the sections in between. This is exemplified by various tributaries of the River Tisza, which flows from Romania to the territory of Hungary, after which its waters return to Romania via Serbia.

Whether or not the actors concerned accurately perceive the significance of this or other comparable peculiarities is of course another matter. Extra-sectoral compensation is therefore generally required to address the structurally unfavorable implications of river-sharing games for healthy water basin governance.

References

Aviram R et al (2014) Desalination as a game-changer in transboundary hydro-politics. Water Policy 16(4):609–624

Barna B (2022) "Tavi bénulásban" elpusztult madarakat találtak Tiszafüred közelében. Egri ügyek. https://egriugyek.hu/kornyek-ugye/tiszafured-tavi-benulas-botulizmus-madarak-fotok. Accessed 23 Sept 2022

Bergmann SM et al (2020) Koi herpesvirus (KHV) and KHV disease (KHVD): a recently updated overview. J Appl Microbiol 129(1):98–103

Berkman DS et al (2002) Effects of stunting, diarrhoeal disease, and parasitic infection during infancy on cognition in late childhood: a follow-up study. Lancet 359(9306):564–571

Blake BE, Fenton SE (2020) Early life exposure to per- and polyfluoroalkyl substances (PFAS) and latent health outcomes: a review including the placenta as a target tissue and possible driver of peri- and postnatal effects. Toxicology 443:152565

Blasco-Costa I et al (2013) Upstream-downstream gradient in infection levels by fish parasites: a common river pattern? Parasitology 140(2):266–274

Bozzay B (2022) Sorra száradnak ki a kutak az országban, és nem egyszerű újakat fúrni helyettük. Telex. https://telex.hu/tudomany/2022/08/11/viz-aszaly-vizgazdalkodas-kut-kutfuras-szarazsag. Accessed 11 Sept 2022

Bush S et al (2018) Cross-border issues: an important component of onchocerciasis elimination programmes. Int Health 10(Suppl. 1):i54–i59

Damewood K (2013) Emerging contaminants: Ag runoff poses health risks. Food Safety News. https://www.foodsafetynews.com/2013/10/emerging-contaminants-potential-health-risks-from-agricultural-runoff. Accessed 11 Sept 2022

Golovina E et al (2021) Sustainable management of transboundary groundwater resources: past and future. Sustainability 13(21):12102

Grantham-McGregor SM, Fernald LC (1997) Nutritional deficiencies and subsequent effects on mental and behavioral development in children. Southeast Asian J Trop Med Public Health 28(Suppl. 2):50–68

Grulovic F (2022) Low water levels on Danube reveal sunken WW2 German warships. Reuters. https://www.reuters.com/world/europe/low-water-levels-danube-reveal-sunken-ww2-german-warships-2022-08-19 Accessed 11 Sept 2022

Gustavsen K et al (2016) Cross-border collaboration for neglected tropical disease efforts: lessons learned from onchocerciasis control and elimination in the Mano River Union (West Africa). Glob Health 12(44):1–5

Gyles C (2016) One medicine, one health, one world. The Canadian Veterinary Journal [La revue veterinaire canadienne] 57(4):345–346

Jin Y et al (2020) Detection of koi herpesvirus (KHV) and carp oedema virus (CEV) in invasive round goby, *Neogobius melanostomus pallas*, 1814, from Poland and Germany. J Vet Res 64(2):247–251

Jonkers ART et al (2010) Epidemics and control strategies for diseases of farmed salmonids: a parameter study. Epidemics 2(4):195–206

Jutla A et al (2013) Environmental factors influencing epidemic cholera. Am J Trop Med Hyg 89(3):597–607

Kibret S et al (2019) Malaria around large dams in Africa: effect of environmental and transmission endemicity factors. Malar J 18(303):1–12

Koch T (2014) 1831: the map that launched the idea of global health. Int J Epidemiol 43(4):1014–1020

Kovács-Bokor É et al (2021) Toxic metal phytoextraction potential and health-risk parameters of some cultivated plants when grown in metal-contaminated river sediment of Danube, near an industrial town. Environ Geochem Health 43:2317–2330

Kvach Y et al (2017) The parasite community of round goby *Neogobius melanostomus* (Pallas, 1814) (*Actinopterygii: Gobiidae*) newly introduced into the upper Elbe. Knowl Manag Aquat Ecosyst 418(19):1–6
Ma H et al (2022) Amplification of downstream flood stage due to damming of fine-grained rivers. Nat Commun 13(3054):1–11
Mackenzie JS, Jeggo M (2019) The One Health approach: why is it so important? Trop Med Infect Dis 4(2):88
Marton P, Szálkai K (2017) Against the current: deconstructing the upstream/downstream binary in Hydropolitical Security Complexes. New Perspectives 25(3):1–27
Moellenkamp S (2007) The "WFD-effect" on upstream-downstream relations in international river basins: insights from the Rhine and the Elbe basins. Hydrol Earth Syst Sci Discuss 4(3):1407–1428
Morris RD (2007) The blue death: disease, disaster, and the water we drink. HarperCollins, New York
Pokorova D et al (2005) Current knowledge on koi herpesvirus (KHV): a review. Vet Med—Czech 50(4):139–147
Rzętała MA (2016) Cadmium contamination of sediments in the water reservoirs in Silesian Upland (southern Poland). J Soils Sediments 16:2458–2470
Sokolow SH et al (2017) Nearly 400 million people are at higher risk of schistosomiasis because dams block the migration of snail-eating river prawns. Philos Trans R Soc B 372(1722):20160127
Steenland K, Winquist A (2021) PFAS and cancer, a scoping review of the epidemiologic evidence. Environ Res 194:110690
Szálkai K (2022) Water-borne diseases. In: Marton P, Romaniuk S (eds) The Palgrave encyclopedia of global security studies [published online, print edition forthcoming]. Palgrave Macmillan, London
Tarrass F, Benjelloun M (2012) The effects of water shortages on health and human development. Perspect Public Health 132(5):240–244
TNMN (2016) Water quality in the Danube River Basin. TNMN Yearbook 2016. International Commission for the Protection of the Danube. https://www.icpdr.org/main/sites/default/files/nodes/documents/tnmn2016.pdf. Accessed 25 Aug 2022
Watson JT et al (2007) Epidemics after natural disasters. Emerg Infect Dis 13(1):1–5
Weidensaul S (2021) A world on the wing: the global odyssey of migratory birds. W. W. Norton, New York
WFD (2000) Water Framework Directive (WFD) 2000/60/EC. Adopted on 23 Oct 2000. https://ec.europa.eu/environment/water/water-framework/index_en.html. Accessed 29 June 2022
Xu L et al (2014) Wet climate and transportation routes accelerate spread of human plague. Proc R Soc B 281(1780):20133159
Yousef SF (2021) Water scarcity and conflict between upstream and downstream riparian countries. Water Econs Policy 7(3):2150012
Zrnčić S et al (2020) Koi herpesvirus and carp edema virus threaten common carp aquaculture in Croatia. J Fish Dis 43(6):673–685

Chapter 13
Transboundary Water Management Among Small Island Developing States: Dynamics of Conflict and Cooperation

Máté Szalai

Abstract The chapter focuses on the role and behavior of small island developing states in issues concerning transboundary water management. Seas and oceans play a crucial role in the politics, economic survival, well-being, and identity of such entities, all of which constitute incentives for cooperation. Despite this generally positive narrative surrounding the collaborative nature of small island developing states, the empirical investigation of key sectors, including fishery, energy and climate, and trade and maritime connectivity, leads to the conclusion that cooperation between such resource-scarce entities is severely limited by several factors. The chapter identifies over-institutionalization, lacking implementation and coordination, as well as political and economic competition between small island developing states as key phenomena in this regard.

Keywords Small island developing states · SIDS · Fishery · Climate change · Energy security · Connectivity

13.1 Introduction

According to the traditional academic perspective, transboundary water management is a predominantly conflictual area of international relations. Access to water sources, the regulation of transboundary aquifers, lakes, or rivers, issues of national water security, or simply the societal role various waters play can easily start, intensify, or contribute to conflicts between states and other political actors, especially if such problems are coupled with power inequality. While cooperation can and most certainly does occur between states, the distribution of costs and information can cause further problems and tensions.

M. Szalai (✉)
Ca' Foscari University, Marie Skłodowska-Curie Cofund Fellow, Venice, Italy
e-mail: mate.szalai@unive.it

K. Szálkai and M. Szalai (eds.), *Theorizing Transboundary Waters in International Relations*, Springer Water, https://doi.org/10.1007/978-3-031-43376-4_13

Small island developing states (SIDS) are usually neglected in the mainstream academic inquiry of transnational water politics,[1] even if they can be useful case studies. Constituting up to 15–25% of all independent states (depending on the definition),[2] their investigation is necessary to understand the tendencies of conflict and cooperation between states regarding transboundary waters. Arguably, most international activities of SIDS are connected to tackling threats and capitalizing on the opportunities arising from shared waters (including seas, oceans, and groundwater systems), which separate them from each other while connecting them geographically, economically, socially, and politically in other ways.

Moreover, the predominantly conflictual understanding of transboundary water management can be juxtaposed with the perception of SIDS as the champions of international cooperation. Generally, small states have been seen by most observers as political entities transcending the realist logic of interstate rivalry and nurturing a collaborative spirit (Thürer 1998). The assumption can be made that such entities are more effective in cooperation in transboundary water management than larger states due to the lack of coercive tools to force their will on others and the preference for rule-based and institutionalized cooperation in international interactions. Naturally, an empirical investigation might contradict this supposition, given the conflictual nature of the field and several other political, economic, or institutional factors that might undermine the efficiency of collaboration.

The aim of this chapter is to assess the different ways small island developing states engage in cooperative or conflictual interactions with each other over the management of issues related to the oceans and seas surrounding them, and to detect the limits of such activities. Naturally, this article cannot provide a complete overview of the cooperation between SIDS either from a regional or a policy perspective; therefore, the research focuses on global or interregional cooperation in three sectors connected to the transboundary nature of seas and oceans: fishery, energy and climate policy, as well as connectivity and trade. This way, the research contributes to the discussion of cooperation and conflict in transboundary water management.

The chapter is divided into three substantive parts. First, the connections in the literature on small states as a case study for transboundary water management are investigated. Second, the basic attributes of SIDS are presented, and the dynamics

[1] One reason is that seas and oceans are not usually considered transboundary waters, even if the most traditional definition set up by the Convention on the Protection and Use of Transboundary Watercourses and International Lakes conceptualises transboundary waters as "any surface or ground waters which mark, cross, or are located on boundaries between two or more states", which can arguably be applicable to seas and oceans. While it is true that the Convention itself does not focus on national and international maritime territories, the argument can be made that the title of the convention does not refer to transboundary waters in general but transboundary watercourses and international lakes in particular. Moreover, the logic regarding the role of transboundary waters as described by Kinga Szálkai (2016) can easily be applied to the case of seas and oceans.

[2] According to the official UN framework, 38 out of its 193 member states are considered small island developing states, 19.6% of all members. The international organisation also recognises 20 additional non-UN member states as SIDS. Given that both the terms "small" and "developed" are debated in the literature, it is impossible to give a proper definition of SIDS, which is why international organisations and researchers tend to provide an exhaustive list of them instead.

of cooperation and conflict are analyzed in connection with transnational waters. Third, three case studies are presented to shed light on the incentives and disincentives for interstate cooperation. Overall, the investigation aims to contribute to the ongoing discussion on the nature of conflict and collaboration in transboundary water management.

13.2 Small States and Transboundary Waters

The analysis of small states' behavior in issues related to transboundary water management has been scarce in the literature, as such entities can be the subject of interesting but challenging case studies. First, even if the assumption can be made that smaller states are more likely to suffer water scarcity (just like in the case of other natural resources), the causal relationship between state size and a lack of water resources is not a direct one. In their empirical investigation, Moncada et al. (2018: 501–502) found only a very weak negative correlation between population size and available cubic meters of water as a ratio to GDP. Naturally, the investigation of other aspects of size (including territory or GDP) might result in different conclusions.

Second, small states are more exposed to external shocks, including environmental and ecological ones. This exposure does not necessarily manifest in more natural or climate crises (although this is the case when it comes to SIDS), but small states tend to have a smaller toolkit to react when such crises occur. Moreover, according to empirical analysis, natural catastrophic events are more likely to lead to a fiscal contraction in small states than in larger ones (Darius et al. 2018: 559). According to the UN (2019: 2), the economic costs of climate change for SIDS (including disasters and other effects) constitute at least 15% of their GDP. At the same time, small states are expected to play an essential role in regional organizations, especially economic ones. These two particularities highlight smaller states' importance in studying transboundary water issues (Koff and Maganda 2015: 233).

Third, it can be argued (Koff 2017; Koff et al. 2020) that the dynamics of cross-border water management reflect local and regional power asymmetries on the one hand, and the strength of shared norms related to transnational cooperation on the other. Therefore, the participation of small states in such frameworks tells us a lot not only about the foreign policy of small states but also about the various characteristics of regional systems. For example, the dynamics between large and small states in transboundary water management are shaped by various aspects of their political culture. Harlan Koff and Carmen Maganda (2015) compare the case of Rio Hondo, which separates Belize and Mexico, and the case of the Moselle, which divides Luxembourg, France, and Germany. They find that even if the legal, institutional, and regulatory framework has been developed much more in the second case, the problems of implementation (especially in terms of social exclusion) have been similar in both cases. Connecting their investigation to the work of Peter Katzenstein, who elaborated on the effects of smallness on economic and political culture, the authors argue that the similarities can be attributed to the propensity of interpersonal relations among political elites, a common characteristic of small state politics, and a feature that was identified among SIDS as well (e.g. Jordan 2007). Among others, this corporatist framework leads to consensus-seeking behavior not only between governments but also between the government and local leaders, enhancing the voice of the latter group in water management. Nevertheless, these local leaders tend to follow their particular interests against their own government, making it challenging to implement environmental policies that serve protective or conservationist aims.

Fourth, by investigating the behavior of small states in transboundary water management, the general wisdom can be questioned about the dominantly positive (and, to some extent, victimizing) narrative about small states. On the one hand, empirical investigation in Central America shows that small states are not necessarily more supportive of regionalism and international regimes in water management than larger ones, questioning the ab ovo cooperative nature of small state foreign policy (Koff et al. 2020). In other cases, when small states do support regional cooperation and the establishment of new legal and regulatory frameworks, they are often ineffective in implementing the rules they helped put in place (Koff and Maganda 2015).

On the other hand, similarly to other sectors, water politics constitute not only a source of threat but also an opportunity for small states to brand themselves. For example, Tuula Honkonen and Annukka Lipponen (2018) show how Finland engaged Sweden, Russia, and Norway to cooperate in transboundary water management. Another policy paper written for the Dutch government by the Clingendael Institute (van Genderen & Rood 2011) argues that the Netherlands should consider water diplomacy as a niche in which it can enhance its influence and image in international society. The authors identify four roles that are available for the Netherlands, showing how the logic of small state foreign policy can effectively lead to enlarged power: as a natural broker in water disputes via the international organizations located in the country; a central hub for international and transnational coordination in international water diplomacy networks; being an enabler by promoting the works of various kinds of non-state actors, e.g. by facilitating consultancy and capacity-building training; and as a norm entrepreneur in terms of developing global public goods in the field of transboundary waters. These specific forms are echoed by the general discussion of the innovative foreign policy strategies of small states (e.g. Chong 2010; Ingebritsen 2006).

13.3 The Case of Small Island Developing States: Incentives for Cooperation and Conflict

Talking about the challenges and opportunities of small states in transboundary water management, arguably the most interesting category includes SIDS, a group of countries that the United Nations designated as a special case of environment and development in 1992 (Binger 2004: 2). While the United Nations recognize 58 SIDS (only 38 of which are UN members,[3] while 20 are non-UN and associate members of the UN)[4] (UN 2022), researchers usually use the category for a far smaller number of states (Roberts 2018: 519). This is mainly due to the questionable smallness (e.g. Cuba), sovereignty (e.g. Guam), or status (e.g. Singapore) of such entities. Moreover, some states on the official list are not even islands (e.g. Belize). Consequently, the actual number of SIDS is debated. In contrast with the general UN designation, UNCTAD has its own list, which comprises 29 countries, whereas the Transboundary Waters Assessment Programme of UNEP only focuses on 42 states. Researchers tend to consider the number of SIDS to be even lower; e.g. John Laing Robert (2018) uses the designation in the case of only 28 countries. While recognizing its limitations, the official UN designation will be used in the framework of this chapter for the sake of simplicity and the applicability of multilateral datasets.

Based on their geographic position, SIDS are traditionally divided into three regional groups, all of which have their own regional cooperative framework (Shultz et al. 2016: 32). The broadest one is the Africa, Indian Ocean, Mediterranean, and the South China Sea cluster, which is institutionalized in the Indian Ocean Commission. The Caribbean SIDS constitute the Caribbean Community, while the SIDS of the Pacific region form the Pacific Islands Forum. Beside these regional forums, the global institutional framework for cooperation for SIDS is the Alliance of Small Island States (ASIS).

SIDS face particular challenges even compared to the broader category of small states in general (Onguglo and Eugui 2018: 531–532). They are fairly isolated from other countries, and their economy is mostly based on the production and export of a few commodities or on providing services (especially in tourism or finances). Consequently, they are highly dependent on other countries and international organizations. Moreover, they are extremely exposed to environmental challenges and catastrophes.

In theory, all of these factors would suggest a greater willingness to engage in cooperation on the side of SIDS, but the picture is much more complicated. Water (i.e. the surrounding seas and oceans) plays multiple political, economic, and societal roles in the development of SIDS, some of which can contribute to or exacerbate interstate conflicts or rivalry. Water is an essential part of the identity of many states. Being surrounded by water creates a self-perception in the form of "islandness", a distinct identity that can be found among "islanders" in parallel with othering "mainlanders" (Nyman 2013: 222–223). Through Elizabeth Nyman's concept of

[3] Antigua and Barbuda, Bahamas, Bahrain, Barbados, Belize, Cabo Verde, Comoros, Cuba, Dominica, Dominican Republic, Fiji, Grenada, Guinea-Bissau, Guyana, Haiti, Jamaica, Kiribati, Maldives, Marshall Islands, Mauritius, Micronesia, Nauru, Palau, Papua New Guinea, Saint Kitts and Nevis, Saint Lucia, Saint Vincent and the Grenadines, Samoa, Sao Tomé and Principe, Seychelles, Singapore, Solomon Islands, Suriname, Timor-Leste, Tonga, Trinidad and Tobago, Tuvalu, Vanuatu.

[4] American Samoa, Anguilla, Aruba, Bermuda, British Virgin Islands, Cayman Islands, Commonwealth of Northern Marianas, Cook Islands, Curacao, French Polynesia, Guadeloupe, Guam, Martinique, Montserrat, New Caledonia, Niue, Puerto Rico, Sint Maarten, Turks and Caicos Islands, U.S. Virgin Islands.

maritime territoriality (Nyman 2013: 222–223), transnational seas and oceans do not separate islands from each other but rather connect them. The islandness identity of many SIDS is reinforced by the already mentioned institutional frameworks for the cooperation of SIDS (primarily ASIS), and it adds another layer to every transnational issue related to seas and oceans, especially if it is connected to defending national identity and sovereignty.

Moreover, water is also the primary source of economic activity for SIDS (Ongluglo and Euguni 2018: 531–532). Most of them have vast areas of territorial water and exclusive economic zones (EEZs), which can be much larger than their actual land territory. In fact, approximately one-third of all seas and oceans belongs to the exclusive economic zones of SIDS (Ongluglo and Euguni 2018: 531–532). Consequently, every economic discussion related to seas and oceans can be regarded as vital for the survival and economic well-being of SIDS, which might foster zero-sum thinking.

Naturally, not all SIDS are directly dependent on the exploitation of their oceanic or natural space. While, for example, Nauru, Papua New Guinea, or Trinidad and Tobago rely profoundly on the export of their natural gas or fish (among others), many SIDS have a service-oriented economy, specializing mostly in tourism or finances (UNCTAD 2014: 5). Using the conceptualization of the Indian Ocean Rim Association, three different roles played by oceanic space in national economies can be differentiated (Voyer et al. 2018: 2):

- oceanic economy: a segment of the economy that is dependent on the ocean, especially in terms of resources needed for production;
- coastal economy: all economic activity that is connected geographically to the coast or coastal regions;
- blue economy: a "much wider and inclusive" term that includes all activities that are ocean-related and which directly or indirectly support activities of these economic sectors (Mohanty et al. 2015: 9).

Using this conceptualization, if a country produces and exports natural gas from its EEZ, it is considered a part of the oceanic economy. Tourism is more likely to be part of the coastal economy, while blue economy also includes the transportation industry and logistics. While the role of transboundary waters is different in the three different kinds of economic activity, all of them can spur competition among SIDS, undermining cooperation efforts.

From a different perspective, the massive amount of water surrounding SIDS and the related transnational issues represent the main source of security and threat to their survival and prosperity. While maritime security is poorly defined in the literature, it plays an important role for SIDS and island states in general (Voyer et al. 2018: 2–4). Bueger (2015) argues that the concept comprises four major elements: sea power (the ability of authorities to protect the state and its interests, including defending sea lanes, claimed maritime zones, and boundaries), marine safety (the security of ships and maritime installations), economic development (the ability of authorities to enforce laws that regulate legal economic activities), and human security (ensuring food security, tackling human trafficking, etc.). Maritime security can be threatened by other states (including larger and smaller ones), but in practice, extreme weather conditions, risks related to climate change, piracy, as well as illegal networks engaged in illicit trade constitute more acute problems for island states in general. For SIDS, these questions are especially important—if maritime security is not maintained, their survival and connectivity can be jeopardized.

Lastly, seas and oceans also function as the sphere where national borders and thus the limits of national sovereignty are constructed. It is practically where the space belonging to SIDS, to mainland countries, or to everybody meets. Therefore, the limits of sovereignty are not connected to lands, rivers, or other geographic constructs but to abstract laws serving to delimit a naturally undividable space. Since 1982, the United Nations Convention on the Law of the Sea (UNCLOS) has defined the rules of the delimitation of maritime boundaries, but that does not mean that there are no sovereignty struggles among SIDS. In spite of the general assumption that small island states are more peaceful than larger states, empirical investigation might show a different picture. Investigating maritime conflicts in the Western Hemisphere between 1990 and 2001, Elizabeth Nyman (2013) found that there was a bigger chance for both peaceful and non-peaceful resolution of disputes if both participants of the conflict were island states. Islandness had a bigger explanatory value than the nature of their political systems (i.e. whether they are both democracies) or capacities (i.e. whether they are both small states). In practice, this means that the special geopolitical status of island states and the shared responsibility over transnational waters probably leads to more urgency to resolve the conflict and does not affect the way in which the conflict is resolved.

All of these observations indicate that both cooperation and conflict are present in the relations of SIDS in issues related to transboundary waters. Thus, neither the idealistic interpretation of small state behavior nor the rather pessimistic view of scholars focusing on transboundary water management is entirely valid. Therefore, in order to see the cooperative and conflictual tendencies among SIDS, specific cases and sectors should be analyzed separately.

13.4 Cooperation and Conflict in Different Sectors Connected to Transboundary Waters

According to the empirical literature, institutionalized cooperation is the main tool in the foreign policy of SIDS for several reasons. First, from a realist perspective, it is self-evident that such small and weak actors should work together to be more influential in the international community (Sutton and Payne 1993: 197–198). Collectively, SIDS have more than 65 million inhabitants (UN 2019) and a GDP of 103.4 billion USD (UNCTAD 2021: 39), the latter of which is bigger than Ethiopia's GDP but smaller than Slovakia's. In terms of GDP per capita, that makes their collective more developed than the broader group of developing countries, as well as countries in transition (UNCTAD 2021: 39).

Second, in accordance with the liberal tradition of International Relations, tackling common threats and realizing profit from mutual sources constitutes a solid basis for institution and regime building. Having such a huge oceanic space makes SIDS highly interconnected with and interdependent on both each other and international waters. Beside facing common challenges, more and more attention is given to opportunities for SIDS, which represent a bigger incentive for institutionalized cooperation. The economic space approach (Ongluglo and Euguni 2018: 533–534) developed by UNCTAD is one narrative that tries to move the attention away from the limitations, toward the possibilities. Instead of viewing SIDS as small island states, such entities can be regarded as "large ocean states" with vast resources, and tools in their hand. The economic space approach advocates for the "co-management and co-benefit sharing of joint maritime resources" and for the combination of national exclusive economic zones into regionally integrated, collectively available spaces (Onguglo and Euguni 2018: 533).

The economic space approach is applicable mostly if the exclusive economic zones of SIDS are directly connected; none of the stakeholder countries have enough financial or human capital for resource exploitation; the shared economic space of multiple countries is affected by migratory species flows or seaways; the sustainability of the management and exploitation of living and non-living resources is threatened; the national capacity of stakeholder countries is limited to implement or monitor policies; or if connections between national authorities are non-existent or minimal. In general, cooperation among SIDS is conducted in four different ways (Sutton and Payne 1993: 197–199): joint representation, foreign policy coordination at the time of a crisis, common regional projection in international bodies, and interregional cooperation. The dynamics between these four methods vary in different fields of cooperation, therefore, a sectoral approach is necessary to evaluate such cooperation formats.

While cooperation remains the norm in transboundary water-related issues among SIDS, the investigation of specific sectors shows that the effectiveness of this norm is heavily limited by various factors and is often undermined by rivalry or economic and political rationale. In the following pages, case studies of different sectors are presented, each of which demonstrates specific dynamics that undermine collaboration in transboundary water management.

13.4.1 Fishery and Over-Institutionalization

Cooperation in fishery represents the primary sector where the economic space approach can be observed in practice. By 2014, UNCTAD (2014: 8) estimated that there were at least 50 fisheries agreements with the participation of SIDS, mostly focusing on the conservation, management, and development of fisheries. Nevertheless, most of these agreements aimed solely at fostering exploitation with a minimal level of cooperation among states. According to critical voices (UNCTAD 2014: 8), such agreements do not take the question of illegal, unreported, and undocumented fishing activities into account, especially in connection with the proper implementation of the Convention on International Trade in Endangered Species of Wild Fauna and Flora, and with their explicit focus on production, they create an environment in which the recovery of stocks is questionable. Moreover, weak institutions and lacking transparency do not help in the evaluation and implementation of the fishery regimes.

Beside these problems, fishery agreements clearly show that cooperation can be hampered by too many different formats and legal bases. Instead of focusing on the implementation of existing documents, SIDS tend to forge new ones with a partly similar scope and jurisdiction. Such proliferation of agreements in the field has not resulted in the emergence of an international cooperation regime, rather a collective of overlapping and non-transparent rules and regulations.

In order to promote coherence and information-sharing, the Food and Agriculture Organization (FAO) has created the Regional Fishery Body Secretariat Network (RSN) (FAO 2022a). This network includes various international bodies and secretariats dealing with the regulation of fishery in national and international waters (FAO 2022b). In the vocabulary of the FAO, the RSN facilitates the exchange of information and cooperation between regional fishery bodies (RFBs), namely "intergovernmental bodies through which States cooperate on the management of fisheries in specific regions" (FAO 2022b). The collaboration is hampered to some extent by the fact that RFBs are very different from one another in multiple respects:

nature:	regional fishery bodies include both organizations and agreements;
mandate:	some RFBs serve merely as a forum for discussion, while others have the capacity to adopt binding measures (these are called regional fisheries management organizations or agreements, RFMO/As);
jurisdiction:	some RFBs only cover waters under national jurisdictions that are directly connected to each other, while others refer to areas beyond national jurisdiction; and
global embeddedness:	some RFBs have been created in the FAO framework, while others are outside of global institutions.

FAO (2022c) recognizes 55 RFBs altogether (Table 13.1). Most of them (44) only deal with one geographic region or ocean, while 11 have a broader scope. Most RFBs have been created in the Atlantic Ocean and the Pacific Ocean.

Beside the multitude of fishing agreements, another problem that limits the effectiveness of the cooperation between SIDS is the fact that they are excluded from the relative majority of such documents. Twenty-seven were created only by larger states and small continental states, while 23 have SIDS and larger states as members. Only one RFB, the Caribbean Regional Fisheries Mechanism, has only SIDS (Table 13.2).

In conclusion, cooperation dominates competition in the fishing industry, but this does not mean that interstate collaboration is completely effective. Over-institutionalization and overlapping jurisdictions lead to a lack of transparency and a diminished capability of stakeholders (states, NGOs, companies, and international

Table 13.1 Geographic distribution of RFBs

Region	Regional Fishery Bodies (RFBs)
Global and trans-ocean (5)	ACAP, CCAMLR, CCSBT, IWC, OSPESCA
Pacific Ocean (13)	APFIC, CCBSP, FFA, IATTC, IPHC, NPAFC, NPFC, PICES, PSC, SEAFDEC, SPC, SPRFMO, WCPFC
The Mediterranean and the Black Sea (1)	GFCM
Indian Ocean (5)	BOBP-IGO, IOTC, RECOFI, SIOFA, SWIOFC
Atlantic Ocean (14)	CECAF COMHAFAT, COREP, CRFM, CTMFM, FCWC, ICCAT, NAFO, NAMMCO, NASCO, NEAFC, SEAFO, SRFC, WECAFC
Continent-related (11)	APFIC, CACFish, LCBC, CIFAA, COPPESAALC, EIFAAC, LTA, LVFO, NACA, MRC, RAA
Other (6)	ACFR, CPPS, CWP, FIRMS, GESAMP, ICES

Source FAO (2022c)

Table 13.2 Membership of RFBs

Members	Regional Fishery Bodies (RFBs)
SIDS (1)	SIDS (1)
SIDS and larger states (23)	IWC, OSPESCA, APFIC, FFA, IATTC, NPFC, SPC, SPRFMO, WCPFC, BOBP-IGO, IOTC, RECOFI, SIOFA, SWIOFC, CECAF, COMHAFAT, COREP, ICCAT, SRFC, WECAFC, APFIC, COPPESAALC, RAA,
Larger states (27)	ACAP, CCAMLR, CCSBT, CCBSP, IPHC, NPAFC, PICES, PSC, SEAFDEC, GFCM, CTMFM, FCWC, NAFO, NAMMCO, NASCO, NEAFC, SEAFO, CACFish, CBLT, CIFAA, EIFAAC, LTA, LVFO, NACA, MRC, CPPS, ICES
Other (4)	ACFR, CWP, FIRMS, GESAMP

Source FAO (2022c)

organizations) to regulate fishery, as well as to the marginalization of SIDS by larger states and companies (Guillotreau et al. 2012). Empirically it is clear that the current state of international fishery regulations is not developed enough to avoid an increase in interstate conflicts (Spijkers et al. 2021), especially given the effects of climate change and illegal fishing. The lack of a clear regulatory framework and the presence of unsustainable fishing practices will likely contribute to more conflicts among SIDS in the future.

13.4.2 Lacking Implementation and Coordination in the Field of Energy and Climate Change

Traditional energy products are usually scarcely available for SIDS, which is one reason why energy security is a major concern for them. SIDS generally face serious challenges in this field—they are extremely dependent on fuel imports, and their electricity networks are not connected either to each other or to other continental networks (Dornan 2018: 317). This directly leads to financial pressure (especially since the 1970s), supply risks, and indirect climate costs (Atteridge and Savvidou 2019: 1; Binger 2004). This is only one reason why the issue of climate change is directly connected to energy security in SIDS, and the triple interlinking threats of natural disasters, climate change, and sea level rise (Shultz et al. 2016:43) pose additional direct threats to many SIDS.

Moreover, the development of renewable energy technologies is widely considered the most important way for SIDS to better develop their infrastructure of marine energy sources, mostly from three sources: offshore wind turbines, generation from waves and submarine geothermal sources, and the usage of algae as biomass for energy production (UNCTAD 2014: 10).

Cooperation among SIDS has accelerated in the past decade. The flagship of this collaboration has been the so-called Samoa Pathway ("SIDS Accelerated Modalities of Action"), a detailed framework for action published at the 3rd International Conference on SIDS in 2014. This wide-ranging document was later strengthened by its inclusion in UN General Assembly Resolution 69/15 in November 2014 (UN General Assembly 2014). It builds on two other key documents passed since 1990, the Barbados Program of Action (1994) and the Mauritius Strategy of Implementation (2005) (Commonwealth Report 2015: 2).

The Samoa Pathway, which refers to various dimensions of sustainable development, mentions affordable and modern energy services, renewable energy, and energy efficiency as shared aspirations of SIDS, echoing many of the national development plans (Atteridge and Savvidou 2019: 2). When it comes to national and collective action, Sect. 12 of the Samoa Pathway emphasizes that any kind of international partnership should be based on the values of national ownership, mutual trust, transparency, and accountability.

Possibly the most tangible product of the Samoa Pathway has been the creation of a new UN-recognized international organization, SIDS DOCK, a cooperation format of small island and developing states for "addressing climate change, resilience, and energy security in small islands" (SIDS DOCK 2022). The organization cooperates with the International Renewable Energy Agency (IRENA) in several projects, including the SIDS Lighthouses Initiative, the Global Geothermal Alliance, and the Island Women Open Network (IRENA 2020). Another interesting form of transnational cooperation was set up in 2021 between SIDS DOCK and the British company Global OTEC Resources (Offshore Energy 2021a), which constructs power plants based on Ocean Thermal Energy Conversion (OTEC), i.e. the conversion of the solar heat energy stored in the oceans. At the end of 2021, Sao Tomé and Príncipe announced the construction of their first OTEC power plant in the framework of SIDS DOCK and Global OTEC, to be financed by the United Nations Industrial Development Organization (Offshore Energy 2021b).

In 2019, a review process took place regarding the implementation of the Samoa Pathway, resulting in a new UN General Assembly Resolution (A/74/L.3). This process was extremely important given the criticism toward small states lacking sufficient feedback mechanisms when it comes to transboundary water cooperation. Accordingly, in Sect. 28, the resolution recognizes that SIDS need "improved data collection and statistical analysis" to "better plan, monitor and evaluate the implementation of" the Samoa Pathway (UN General Assembly 2019). On a similar note, Sect. 30 calls for relevant national institutions to learn more effectively from each other and share best practices.

On the other hand, the document acknowledges the efforts of small island developing countries (especially in terms of advancements in social inclusion), and it also mentions a few specific initiatives, especially those conducted by the International Renewable Energy Agency and SIDS DOCK. Nevertheless, the document (beside other sources) calls for further improvement regarding women's empowerment and inclusion in sustainable development, tackling challenges in accessing affordable financing, and exploring unorthodox financial instruments to help SIDS (e.g. the introduction of blue or green bonds or debt for development swaps), and their better integration in multilateral trade platforms. Another report (Commonwealth Foundation 2015: 6) on the Samoa Pathway highlights the fact that better cooperation between the governments of small island states and their civil societies might be needed in both implementation, cooperation, and following up on the promises of international organizations. A research paper focusing on climate-related information and knowledge management in Fiji, Tonga, and Vanuatu found that there is no comprehensive framework to guide the implementation of national and collective climate action, while data sharing both domestically and internationally is minimal or non-existent (Mackay 2019: 131–132).

In conclusion, the presented case study shows that even though there is a clear and general political will to cooperate among SIDS, the lack of coordination and implementation capacity undermines achieving success. While limited resources do play an explanatory role in the problem, there are other narratives to interpret the causes behind these shortcomings. First, SIDS are very different from one another in many respects (Dornan 2018: 317–320). Their energy market is small in global comparison, but relative to each other, the differences are enormous, exemplified by the cases of Cuba, Bahrain, or Tuvalu. Moreover, social access to energy and electricity varies highly (correlating with the general level of economic development), as does the potential for renewable energy development. Second, the structure of the national energy markets and the related policies do not incentivize cooperation, either (Dornan 2018: 317–320). Most SIDS continue to use a sort of public monopoly model, and reform initiatives have only been observable recently. These policies, nonetheless, have primarily introduced highly ambitious renewable energy targets, for which SIDS will need additional funding (including development assistance, climate change funds, or private investments). This necessarily creates competition among SIDS. Innovative, cost-friendly solutions do exist in SIDS, although these are usually local methods without an international component. For example, Jamaica is trying to overcome its freshwater scarcity by harvesting rainwater locally (Waite 2012), even if international cooperation regarding information sharing and finances would be beneficial.

Third, the spread and development of ocean energy technologies has been slower than expected in the past decade (IRENA 2022). More private funding is needed, which is beyond the individual or collective reach of SIDS. Fourth, as was described in a Commonwealth Foundation Report (2015: 6) and other documents (Mackay 2019), institutional cooperation among the national authorities of SIDS at the local, regional, and international levels is often inadequate. In spite of the multitude of existing frameworks, national actions are not coordinated due to institutional shortcomings. The study of Mackay points out one best practice in the Pacific, the South Pacific Regional Environmental Program (SPREP), which has proven to be highly effective not only in coordination but also in channeling international financial support for SIDS.

13.4.3 Intense Competition in Trade and Maritime Connectivity

Trade and maritime connectivity is a vital sector for SIDS, which is why cooperation and conflict are present in this area simultaneously. Due to their remote geopolitical locations, the development of maritime trade and the transportation infrastructure is especially important for SIDS. Without stable and reliable linkages between SIDS and the rest of the world, island states are not only deprived of their economic possibilities, they also lose political and social connectivity. Statistically, lacking direct maritime connection is associated with as much as 40% lower bilateral exports (Fugazza and Hoffmann 2017). In consequence, the fact that more than half (29) of the 50 least interconnected countries are SIDS is a huge obstacle to their development (UNCTAD 2021: 55).

In practice, boosting the infrastructure poses many challenges (UNCTAD 2014: 13; UNCTAD 2021: 55–56). Beside diseconomies of scale and lacking competitiveness, natural disasters, climate change risks, fragile ecosystems, interdependences between island states, and lacking institutional potential and attractiveness for investment easily hamper SIDS' initiative to establish more maritime routes.

Lacking infrastructure is not the only difficulty SIDS are facing in trade policy. As most small island states were under colonial rule before independence, they continued to have a special trade relationship with their former ruler (Binger 2004: 1–2). This can be considered a relatively beneficial situation, as it protected the special status of SIDS with large national economies. Nevertheless, the emergence of the World Trade Organisation (WTO) regime in the 1990s made it impossible to sustain such preferential relationships except for an agreed transitional period. After that, SIDS had to compete with bigger producers with better connectivity, as well as with each other (Binger 2004: 1–2).

It is not surprising that transportation connectivity is one of the key areas where cooperation is the biggest among SIDS. The review of the already mentioned Samoa Pathway also emphasizes that cooperation in this field has been commendable (UN General Assembly 2019: Sec.25). Nevertheless, SIDS are also engaged in natural competition in developing their maritime connectivity. Beside trade, this rivalry can be observed in tourism, a sector that accounts for more than 20% of the GDP for two out of three SIDS (OECD 2018: 31), and where competition has become global (Spencer 2019: 116). In such market conditions, it would be rational for regional SIDS to market themselves "under one umbrella", but such cooperation is often overshadowed by cutthroat competition (Barrett et al. 2014: 117; Spencer 2019: 119). Beside general interstate rivalry, core-periphery dynamics also shape relations between islands, even if they belong to the same state (e.g. Trinidad and Tobago) (Jordan 2007).

When it comes to connectivity and transportation infrastructure, Jamaica, Mauritius, and the Bahamas are far ahead compared to most other SIDS, as they have been performing almost constant development in this field since the mid-2000s (UNCTAD 2021: 55–56). SIDS host some well-connected ports in Kingston (Jamaica), Caucedo (Dominican Republic), and Port Louis (Mauritius), while Port Mathurin of Mauritius, Palmeira of Cabo Verde, and Luganville of Vanuatu are the least integrated into shipping routes.

Another indicator of the relative connectivity and competitiveness of SIDS is the number of shipping service companies registered nationally (UNCTAD 2021: 57). From this perspective, a general decline has been observed in most cases during the last two decades, primarily due to the global economic crisis of 2008 and the global effects of the Covid-19 crisis. While Trinidad and Tobago and Jamaica are at the top of the list, they too have fewer companies working in the sector than in the mid-2000s. Despite a few exceptions (e.g. the Marshall Islands), most SIDS (including Antigua and Barbuda or Mauritius) have suffered a similar decline, while Micronesia or Saint Kitts and Nevis have managed to keep their numbers relatively steady. Beside supporting the creation of national companies, an alternative tool in the hands of the governments of SIDS is to incentivize ship registration financially and administratively. Some SIDS like the Marshall Islands, Tuvalu, or Vanuatu are among the nations hosting the highest number of merchant ships (UNCTAD 2014: 13).

All in all, trade and connectivity are the areas where rivalry and competition might be the most visible between SIDS, which undermines cooperation and may potentially lead to conflict. Moreover, this case study also shows that the nature of the relationships among SIDS is highly affected by changes in global trade regulations, consequently, it is not only individual states that are exposed to developments beyond their control, so are their relationships.

13.5 Discussion and Conclusions

A review of the interactions of small island developing states in issues related to transboundary water shows that cooperation remains the only viable option for them in most issues connected to seas and oceans. Beside the more collaborative norms incentivized by the identity of smallness and weakness, practical notions also contribute to the deepening of different frameworks of collaboration among SIDS. The identity of islandness, exposure to threats related to climate change, and economic interdependence are the main reasons why SIDS might be better at cooperation in transboundary water management than larger states.

Nevertheless, the effectiveness and scope of this cooperation is limited by several factors that play different roles in different sectors. In terms of the fishery sector, too many institutional agreements and forums undermine transparency and information sharing. In matters related to energy and climate, a lack of implementation, institutional shortcomings, and barely existing feedback mechanisms undermine effectiveness. When it comes to trade and connectivity, the often overlooked but natural competition among SIDS hampers collective action.

Overall, the analysis of small island developing states highlights four main conclusions regarding the dynamics of cooperation and conflict in transboundary water management and related activities. First, although it seems that while neither the optimistic expectations of small state studies nor the pessimistic attitude of the transboundary water management literature is generally correct, the former has larger explanatory value. Smallness and exposure do encourage SIDS to cooperate more intensively, even in questions that are vital to their survival and where zero-sum logic could arise.

Second, beside size and a lack of political power, weak institutional capacities play a dominant role in understanding interstate interactions (e.g. King and Robinson 2014: 249). Even if there is mutual interest and existing political will to cooperate in a complex issue related to transboundary waters, a lack of experience and administrative capacities can undermine or slow down any meaningful progress. This factor is especially crucial given the fact that the performance of state institutions varies extensively among SIDS, but collectively they tend to perform slightly weaker than aid-receiving countries in general. They are also overrepresented in the fragile state category (McGillivray et al. 2010: 820–821).

Third, while the main dividing line lies between smaller and larger states, inequality surfaces among SIDS as well. While they share vital common interests, their foreign policy is dominated by attracting foreign capital and subsidies from the same sources as other SIDS. While SIDS do cooperate to get additional concessional financial sources, 79% of this income is received on a bilateral basis (OECD 2018: 16). Consequently, current trends show that the distribution of this source of income is highly uneven both in regional and national comparison (OECD 2018: 55–57). Between 2012 and 2015, seven SIDS[5] were the recipients of 64% of all concessional

[5] Haiti, Papua New Guinea, Dominican Republic, Timor-Leste, Cabo Verde, Solomon Islands, Cuba.

subsidies. In comparison, the least subsidized SIDS accounted for only 4%. Such inequality almost necessarily leads to two outcomes, both of which limit cooperation: competition on the one hand and different opportunities for development on the other.

Fourth, as was mentioned in the case study on trade and connectivity, the dynamics of the relationship among SIDS can easily transform if international trade regulations change. Their general collaborative attitude might be an attribute not predetermined by their size or geo-economic situation but a series of other circumstances that are flexible and beyond their control. Based on this observation, the future of the relations among small island developing states seems quite grim, especially given the effects of climate change and great power interests.

References

Alesina A, Enrico S (2005) The size of nations. MIT Press, Cambridge

Atteridge A, Savvidou G (2019) Development aid for energy in small island developing states. Energy, Sustain, Soc 9(10):1–10

Barrett K et al (2014) The debt experience of SIDS in the Atlantic, Indian Ocean, Mediterranean, and South China Sea. In: King D, Tennant DF (eds) Debt and development in small island developing states. Palgrave MacMillan, Cham, pp 91–134

Binger A (2004) Potential and future prospects for ocean thermal energy conversion (OTEC) in small islands developing states (SIDS). Tethys. https://tethys.pnnl.gov/sites/default/files/publications/Binger-2004.pdf. Accessed 17 May 2022

Bueger C (2015) What is maritime security? Mar Policy 53(C):159–164

Chong A (2010) Small state soft power strategies: virtual enlargement in the cases of the Vatican City State and Singapore. Camb Rev Int Aff 23(3):383–405

Commonwealth Foundation (2015) The SAMOA pathway: recommendations from Commonwealth civil society. https://commonwealthfoundation.com/wp-content/uploads/2015/08/Commonwealth-Insights_SAMOA-Pathway.pdf. Accessed 17 May 2022

Convention on the Protection and Use of Transboundary Watercourses and International Lakes. UN. https://unece.org/DAM/env/water/publications/WAT_Text/ECE_MP.WAT_41.pdf. Accessed 17 May 2022

Darius R et al (2018) The fiscal consequences of natural disasters: does country size matter? In: Briguglio L (ed) Handbook of small states: economic, social, and environmental issues. Routledge, London-New York, pp 550–572

Dornan M (2018) When international 'best practice' is not: power sector reform in small island states. In: Briguglio L (ed) Handbook of small states: economic, social, and environmental issues. Routledge, London-New York, pp 313–325.

East MA (1973) Size and foreign policy behavior: a test of two models. World Polit 25(4):556–576

FAO (2022a) Regional Fishery Body Secretariats Network (RSN). https://www.fao.org/fishery/en/rsn/en. Accessed 17 May 2022

FAO (2022b) Regional bodies involved in the management of deep-sea fisheries. https://www.fao.org/in-action/vulnerable-marine-ecosystems/background/regional-fishery-bodies. Accessed 17 May 2022

FAO (2022c) Regional fishery bodies. https://www.fao.org/fishery/en/rfb. Accessed 17 May 2022

Fox AB (1959) The power of small states: diplomacy in World War II. University of Chicago Press, Chicago

Fugazza M, Hoffmann J (2017) Liner shipping connectivity as determinant of trade. J Shipp Trade 2(1):1–18
Garai N et al (2017) A kisállamok lehetőségei és kihívásai a 21. században. In: Garai N, Koncz-Kiss J, Szalai M (eds) A kisállamok lehetőségei és kihívásai a 21. században. Institute for Foreign Affairs and Trade, Budapest, pp 3–60
Guillotreau P et al (2012) Vulnerability of small island fishery economies to climate and institutional changes. Current Opinion in Environ Sustain 4(3):287–291
Handel M (2006) Weak states in the international system. In: Ingebritsen C, Neumann IB, Gstöhl S, Beyer J (eds.), Small states in International Relations. University of Iceland Press, pp 149–153
Hey JAK (ed) (2003) Small states in world politics: explaining foreign policy behavior. Lynne Rienner Publishers, Boulder
Honkonen T, Lipponen A (2018) Finland's cooperation in managing transboundary waters and the UNECE principles for effective joint bodies: value for water diplomacy? J Hydrol 567(December):320–331
Ingebritsen C (2006) Norm entrepreneurs: Scandinavia's role in world politics. In: Ingebritsen C, Neumann IB, Gstöhl S, Beyer J (eds.), Small states in International Relations. University of Iceland Press, pp 273–286
IRENA (2020) IRENA and SIDS DOCK sign agreement to reinforce small island climate resilience. https://www.irena.org/newsroom/pressreleases/2020/Nov/IRENA-and-SIDS-DOCK-Sign-Agreement-to-Reinforce-Small-Island-Climate-Resilience. Accessed 17 May 2022
IRENA (2022) Accelerating the development of OTEC in small island developing states meeting. https://www.irena.org/events/2022/Feb/Accelerating-the-development-of-OTEC-in-Small-Island-Developing-States-meeting. Accessed 17 May 2022
Jordan L (2007) Interorganisational relationships in small twin-island developing states in the Caribbean—The role of the internal core-periphery model: the case of Trinidad and Tobago. Curr Issue Tour 10(1):1–32
Katzenstein P (2003) Small states and small states revisited. New Polit Econ 8(1):9–30
Keohane RO (2006) Lilliputians' dilemmas: Small states in international politics. In: Ingebritsen C, Neumann IB, Gstöhl S, Beyer J (eds), Small states in International Relations. University of Iceland Press, pp 55–77
King D, Robinson M (2014) An agenda for debt sustainability in SIDS. In: King D, Tennant DF (eds) Dept and development in small island developing states. Palgrave MacMillan, Cham, pp 239–250
Koff H (2017) «Paz ambiental transfronteriza» como interacción de normas regionales y poder local: Lecciones desde análisis y debates transregionales sobre seguridad hídrica. Revista De Paz y Conflictos 9(2):19–42
Koff H et al (2020) Transboundary water diplomacy among small states: a giant dilemma for Central American regionalism. Water Inter 45(4):275–291
Koff H, Maganda C (2015) Against the current: transboundary water management in small states on two continents. Water Inter 40(2):231–250
Mackay S et al (2019) Overcoming barriers to climate change information management in small island developing states: lessons from Pacific SIDS. Clim 19(1):125–138
McGillivray M et al (2010) Vulnerability, trade, financial flows and state failure in small island developing states. J Dev Stud 46(5):815–827
McGowan PJ, Gottwald K (1975) Small state foreign policies: a comparative study of participation, conflict, and political and economic dependence in Black Africa. Int Stud Quart 19(4):469–500
Mohanty SK et al (2015) Prospects of blue economy in the Indian Ocean. Research and information system for developing countries. https://www.ris.org.in/sites/default/files/Final_Blue_Economy_Report_2015-Website.pdf. Accessed 17 May 2022
Moncada S et al (2018) Environmental economics: spacial considerations for small states. In: Briguglio L (ed) Handbook of small states: economic, social, and environmental issues. Routledge, London-New York, pp 493–516.

Moses JW (2000) OPEN states in the global economy: the political economy of small-state macroeconomic management. Palgrave Macmillan, Basingstoke

Neumann IB, Gstöhl S (2006) Introduction. Lilliputians in Gulliver's world? In: Ingebritsen C, Neumann IB, Gstöhl S, Beyer J (eds.), Small states in International Relations. University of Iceland Press, pp 3–39

Nyman E (2013) Island exceptionalism and international maritime conflicts. Prof Geogr 65(2):221–229

OECD (2018) Making development co-operation work for Small Island Developing States. https://www.oecd-ilibrary.org/development/making-development-co-operation-work-for-small-island-developing-states_9789264287648-en. Accessed 17 May 2022

Offshore Energy (2021a) Alliance formed to bring floating OTEC to small islands. https://www.offshore-energy.biz/sids-dock-global-otec-resources-join-forces-to-bring-floating-otec-to-small-islands. Accessed 17 May 2022

Offshore Energy (2021b) Deal in place for Sao Tomé and Príncipe's first OTEC power plant. https://www.offshore-energy.biz/deal-in-place-for-sao-tome-and-principes-first-otec-power-plant. Accessed 17 May 2022

Oguglo BF, Eugui DV (2018) Small island developing states and the ocean economy. In: Briguglio L (ed) Handbook of small states: economic, social, and environmental issues. Routledge, London-New York, pp 530–549

Roberts JL (2018) Small island developing states and Sustainable Development Goals: curse or cure. In: Briguglio L (ed) Handbook of small states: economic, social, and environmental issues. Routledge, London-New York, pp 517–529

Samoa Pathway (2014) https://sustainabledevelopment.un.org/samoapathway.html. Accessed 17 May 2022

Shultz JM et al (2016) Disaster risk reduction and sustainable development for small island developing states. Disaster Health 3(1):32–44

SIDS DOCK (2022) What is SIDS DOCK? https://sidsdock.org/what-is-sids-dock. Accessed 17 May 2022

Spencer A (2019) Travel and tourism in the Caribbean. Challenges and opportunities for small island developing states. Palgrave Macmillan, Cham

Spijkers J et al (2021) Identifying predictors of international fishery conflict. Fish Fish 22(4):834–850

Stefano M et al (2018) Environmental economics: special considerations for small states. In: Briguglio L (ed) Handbook of small states: economic, social, and environmental issues. Routledge, London-New York, pp 493–516

Sutton P, Payne A (1993) Towards a security policy for small island and enclave developing states. The Journal of Commonwealth & Comparative Politics 31(2):193–200

Szálkai K (2016) A vízmegosztás kérdése a nemzetközi kapcsolatok elméletében: Vízügyi konfliktusok és együttműködések Közép-Ázsiában a XXI. század elején. Doctoral dissertation, Corvinus University of Budapest, Budapest.

Thürer D (1998) The perception of small states: myth and reality. In: Goetschel L (ed) Small states inside and outside the European Union: interests and policies. Kluwer Academic Publishers, Dordrecht, pp 33–42

UN (2019) SAMOA Pathway high-level midterm review. https://www.un.org/en/summits2019/pdf/SAMOA-Pathway-High-Level-Midterm-Review.pdf. Accessed 17 May 2022

UN (2022) List of SIDS. https://www.un.org/ohrlls/content/list-sids. Accessed 17 May 2022

UN General Assembly (2014) Resolution A/RES/69/15. https://documents-dds-ny.un.org/doc/UNDOC/GEN/N14/628/45/PDF/N1462845.pdf?OpenElement. Accessed 17 May 2022

UN General Assembly (2019) Resolution A/74/L.3. https://www.un.org/ga/search/view_doc.asp?symbol=A/74/L.3&Lang=E. Accessed 17 May 2022

UNCTAD (2014) The oceans economy. Opportunities and challenges for small island developing states. https://unctad.org/system/files/official-document/ditcted2014d5_en.pdf. Accessed 17 May 2022

UNCTAD (2021) Development and globalization: facts and figures 2021. Small Island Developing States. https://unctad.org/system/files/official-document/dgff2021_en.pdf. Accessed 17 May 2022
van Genderen R, Rood J (2011) Water diplomacy: a niche for the Netherlands? Clingendael Institute. https://www.clingendael.org/sites/default/files/pdfs/20111200_cling_report_waterdiplomacy_rgenderen_jrood.pdf. Accessed 17 May 2022
Vital D (1967) The inequality of states: a study of the small power in International Relations. Clarendon Press, Oxford
Voyer M et al (2018) Maritime security and the blue economy: intersections and interdependences in the Indian Ocean. J Indian Ocean RegN 14(1):28–48
Waite M (2012) Climate-change mitigation and adaptation in small island developing states: the case of rainwater harvesting in Jamaica. Sustain: Sci, Pract Policy 8(2):81–87

Part V
Conclusion

Chapter 14
Conclusion: The Role of IR Theories in Transboundary Water Politics

Máté Szalai and Kinga Szálkai

Abstract The concluding chapter attempts to contribute to the ongoing debates in the literature concerning how theories of International Relations can be used in the field of transboundary water politics on the one hand and how case studies related to transboundary waters can contribute to theorizing in International Relations on the other. From these perspectives, the results of the book can lead us to several conclusions and can add value to academic discussions primarily concerning the emergence of mid-level theories, the better management of interdisciplinary research, connecting material and immaterial factors in the analysis, finding and expanding synergy between various subfields of International Relations and adjacent research areas, and the developing the vocabulary of traditional security studies. The chapter ends with the summary of practical implications of the research project which could serve the basis for policy recommendations.

Keywords International relations theory · Transboundary water politics · Hydropolitics · Interdisciplinarity · Security studies

14.1 International Relations and Transboundary Water Politics

The role of International Relations (IR) theories in the field of transboundary water politics (TWP) continues to be a debated one. By contrasting the traditional and current theoretical approaches toward the general dynamics of water-related transnational conflicts and cooperation and investigating the applicability of IR theories in specific case studies, the authors of the present book contribute to the ongoing debate

M. Szalai (✉)
Ca' Foscari University, Marie Skłodowska-Curie Cofund Fellow, Venice, Italy
e-mail: mate.szalai@unive.it

K. Szálkai
Budapest Metropolitan University (METU), Budapest, Hungary
e-mail: kszalkai@metropolitan.hu

K. Szálkai and M. Szalai (eds.), *Theorizing Transboundary Waters in International Relations*, Springer Water, https://doi.org/10.1007/978-3-031-43376-4_14

about identifying how the two fields can interact with each other in a way that is beneficial for both academics and professionals. The underlying question is not only how we can use IR theories to advance discussions related to TWP but also if the analysis of water-related cases can advance IR theories as well.

Naturally, the two research areas have never been completely independent from each other, but traditionally the relationship between them has not been problematized to a great extent. As Kathryn Furlong (2006: 439) puts it, the theoretical underpinnings in TWP "are implicit and unacknowledged". This tendency, according to Furlong, leads to mis-theorization, undue pessimism, and neglecting important aspects. While Furlong is quite pessimistic about the usefulness of the application of IR theories, Jeroen Warner and Mark Zeitoun (2008) challenge her assessment, claiming that mixing IR theories with water scholarship was a "more potent brew". Warner and Zeitoun argue that critical and reflective scholarship is a better fit than the traditional frameworks and schools. Frédéric Julien (2012) echoes their concerns, supporting a constructivist turn in the study of hydropolitics, especially by focusing more on non-rationalist theoretical frameworks.

The case studies presented in this volume add several viewpoints and concepts to this almost two-decade long debate. Summarizing the findings presented by the authors, our research contributes to five ongoing discussions in the transboundary water literature: contrasting the application of grand theoretical schools and mid-level theories; taking into account the consequences of the interdisciplinarity of the field; connecting material and immaterial factors more organically; finding synergy with IR-adjacent areas of research; surpassing the traditional conflict-cooperation nexus in security studies and water diplomacy; and identifying the practical relevance of the TWP literature.

14.2 Grand Theoretical Schools and the Role of Mid-Level Theories

Regarding the debate as to how traditional schools or the critical literature of IR theory could contribute to the literature on transboundary water politics, one of the major arguments of this book is that mid-level theories can provide a better explanatory framework than grand theories. Instead of applying the general logic and vocabulary of realism, liberalism, constructivism, or any other school, starting the investigation at a lower level of abstraction can be more useful. Similarly to other areas covered by IR, TWP should arguably also focus on eclectic or mid-level theories instead of great schools and the debates between them, concentrating on "what works" (Lake, 2013: 568) and not on what is more compatible with a philosophy.

Naturally, the major theoretical schools of IR can, sometimes, provide an ideal starting point for investigating a specific case study. *Zoltán Vörös* successfully uses the realist notion of hydro-hegemony to interpret power relations between China and other East and Southeast Asian countries. In the investigation of the role of data-sharing in the Helmand River Basin, the framework of *Najibullah Loodin, Gabriel Eckstein, Vijay P. Singh* and *Rosario Sanchez* is largely built on the premises of neoliberalism (even if they alter the relevant theoretical model). For *Bota Sharipova*, social constructivism serves as an ideal framework to track the role of trust, emotions, and beliefs in the Aral Sea Basin, while *Fatine Ezbakhe* even argues that there has been a social constructivist turn in the hydropolitics literature, with the emphasis moving toward analyzing discourses and perceptions. Nevertheless, for most of our authors, mid-level theories like securitization or a mixture of different theoretical logics provide a better analytical framework.

There are several reasons why traditional highly theoretical approaches fail to be completely useful for understanding transboundary water politics. Several authors found that the main problem is the perceived complexity of TWP, which needs a simultaneous geopolitical, economic, social, and ecological analysis that touches upon domestic, international, and transnational viewpoints, affected by the local, state, international, and global levels at the same time. While we agree with Grandi (2020) that this interdisciplinary and transdisciplinary nature is a strength in analyzing hydropolitics, it can also be argued that such a complexity inevitably leads to methodological and conceptual difficulties that need to be recognized. This does not mean that abstract theories should not be applied, only that simplified approaches or exclusive interpretations would only lead to incomplete understandings.

Another reason why a single theoretical framework should not be chosen with the ambition to explain the nature of transboundary water politics is actorpluralism. While all theoretical schools have their preference of actors to work with, in the field of TWP, a clear preference or focus would create a blind spot for the researcher. Therefore, we argue that theoretical approaches that are not or hardly capable of simultaneously analyzing different kinds of actors would be ineffective in interpreting transboundary water politics.

Consequently, multistakeholderism could be one mid-level approach that could be useful in transboundary water politics. Mark Raymond and Laura DeNardis (2015: 573) conceptualize the term as a competitor to multilateralism and define it as "two or more classes of actors engaged in a common governance enterprise concerning issues they regard as public in nature and characterized by polyarchic authority relations constituted by procedural rules". The applicability of this concept is also indicated by the fact that the classes Raymond and DeNardis enlist (states, intergovernmental organizations, firms, and civil society actors) are all very active in transboundary water management and have been mentioned by several of our authors. Although she does not call it multistakeholderism, Offutt (2022: 585) also describes a similar methodology and argues for the investigation of stakeholders at different levels at the same time.

Beside serving as an ideal conceptual framework, multistakeholderism also highlights the importance of authority and authority relations in transboundary water politics. While anarchy has been the typical point of departure in the interpretation of water conflicts (as shown by *Atal Ahmadzai*), shifting the focus from power, vulnerability, and risks (Offutt 2022) to how authority is shared or can be attributed to specific actors or a set of rules or institutions can provide a better explanation for the behavior of various stakeholders. This idea is closely connected to the discussion of the role of international institutions in water diplomacy, as highlighted by *Ayşegül Kibaroğlu*'s chapter.

The applicability and usefulness of the notion of multistakeholderism might be best presented by the chapter of *Kinga Szálkai* and *Mary Durfee*, in their research regarding the renegotiation process of the Columbia River Treaty, they build a conceptual framework based on this term. Their case shows that the viewpoint of multistakeholderism is useful from at least three perspectives in transboundary water politics: to introduce actors of various nature, to interpret the relations and dynamics between them in a given process, and to shed light on the different types of authority they possess.

Besides their study, the research of *Atal Ahmadzai* was also built on the inclusion of actors of various natures, including the Pakistani army and the World Bank with a special emphasis on the transnational relationship between them. By avoiding the traditional state-centric approach, Ahmadzai proved that the simplistic interpretation of interstate conflict and cooperation constitutes a huge blind spot for observers of transboundary water politics.

As a result, the analytical approach of picking one grand theoretical framework in IR as a starting point and expecting case studies to align with the theoretical assumptions seems increasingly obsolete. Several of our researchers find the major traditional schools of the discipline not only compatible in explaining transboundary water issues but showing synergy in their parallel application. In *Sameh Al-Muqdadi*'s case study, the respective logics of realism, liberalism, constructivism, and Marxism explain different phases of the same process. While they start their investigation of the nature of data-sharing in the realm of neoliberalism, *Najibullah Loodin* and his colleagues. augment the theory of planned behavior (TPB) model by including rationalist and idealist elements. According to their argumentation, the commitment of a given actor to data sharing is not only shaped by rational calculations and their national interest but also by emotions, history, pride, and national identity, beside tendencies of securitization.

14.3 Consequences of Interdisciplinarity

Such inherent and almost unavoidable interdisciplinarity has far-reaching consequences. On the one hand, researchers in the field have to construct and choose complex methodological frameworks and should not be afraid of modifying existing ones. The need for precise definitions and thorough conceptualizations plays an even greater role than in other disciplines. On the other hand, general realist, liberal, or constructivist observations should not be automatically applied to the field, as their relevance could be highly limited. Moreover, importing the debate that take place between or in the framework of major schools of International Relations to the field could also have adverse effects, especially in interaction with each other.

Naturally, such observations are not limited to the field of transboundary politics. Several other fields, including the ever-expanding security studies, as well as environmental or health politics bear very similar attributes. Since assessing the consequences of interdisciplinarity may have been developed more in these fields than in TWP, it would be advisable to learn from these areas as well. For example, John Cairns (2001) has elaborated on ethical issues concerning the creation and management of interdisciplinary research teams in environmental politics, a challenge that is also present in the field in transboundary water politics. One observation that is especially important for us is the danger of the tribalism of different disciplines the representatives of which should collaborate in a project. This can easily lead to isolating mechanisms (including disciplinary jargon, parallel and exclusive meetings) that undermine communication and the efficiency of research. While these challenges can be overcome, special attention should be paid to handling them.

14.4 Connecting Material and Immaterial Factors

The need to create a useful symbiosis of various theoretical viewpoints in analyzing transboundary water politics is also enhanced by the importance of material and immaterial factors. For many observers, it is easy to recognize the importance of the material factors of water politics, including its geopolitical, economic, or ecological aspects. Nevertheless, in alignment with the constructivist turn identified by *Fatine Ezbakhe*, several of our studies show that normative aspects, values, and ideas heavily influence transboundary water politics. *Atal Ahmadzai* talks about the role of nationalistic discourses, which can strengthen the conflictual nature of interactions and can be used by domestic elite circles to centralize decision-making processes. *Máté Szalai* highlights the fact that even for small island developing states, which are possibly the most exposed to the negative effects of the global climate crisis, seas and oceans are not only a sphere of threats or economic possibilities but also an important element in the self-definition of their societies. Therefore, it is not sufficient to talk about national identities in these cases, as the identity of islandness should also be considered. Analyzing the Aral Sea Basin, *Bota Sharipova* highlights the role of trust and mistrust in transboundary water issues, notions that are outside the scope of traditional IR schools. In their endeavor to update the TPB model, *Najibullah Loodin* and his colleagues connect the notions of national interest, security, and rational calculations with emotions, pride, history, and national identity. They find that in practice, all of these factors need to be investigated to understand both an actor's commitment to data sharing and the actual practice of it.

Finally, the study of *Zoltán Vörös* explains how China exercises its hegemonic role through means beyond the more conventional geographic and material power, including bargaining and ideational power. His differentiation of the different types of power related to hydro-hegemony can be useful in other case studies as well.

14.5 Finding Synergy with Other Sub-Disciplines Adjacent to International Relations

While the ultimate aim of our project was to analyze the applicability of IR theories to case studies provided by transboundary water politics and to articulate lessons for both fields, the authors of the book have found intense vertical synergy with other fields. In practice, this means that the interdisciplinary nature of TWP can be further widened by the inclusion of other viewpoints.

Naturally, such endeavors had already characterized the discipline. Among the usual suspects, security studies have turned out to be particularly useful in our project as well. Environmental politics, climate change, or institutionalism are mentioned several times, as are specific notions of international law. The vocabulary of foreign policy analysis is often utilized, although mostly implicitly, similarly to the majority of the literature on transboundary water politics, which suggests that more explicit investigation can be conducted to explore all forms of synergy.

Beside the most common fields that show substantial overlap with transboundary water politics, the authors of our volume have found sub-disciplines that are seldom mentioned. *Péter Marton*'s study shows how linking public health and transboundary waterways can lead to new research results. Similarly, *Máté Szalai*'s investigation tried to reconcile the various viewpoints of small state studies and transboundary water politics. The chapter by *Kinga Szálkai* and *Mary Durfee* highlights the role of indigenous peoples in water-related negotiations, which suggests that postcolonial studies could also represent a novel direction in which transboundary water politics could expand. Szálkai and Durfee also touch upon the notion of justice in international relations, which opens the door for many novel theoretical viewpoints to be utilized, including the English School or international legal studies. They also talk about the notion of justice in international relations, which opens the door for many novel theoretical viewpoints to be utilized, including the English School or international legal studies.

14.6 Security, Water Wars, and Water Diplomacy

Interpreting the nature and dynamics of conflict and cooperation in the field of hydropolitics has become one of the most frequently analyzed issues, possibly due to the implicit dominance of the realist school of IR (Furlong 2006: 439). As Péter Marton and Kinga Szálkai argue (2017: 16–17), this inquiry has been at least partly driven by the growing global concerns about water scarcity, and the evolving political agenda of climate change. These processes coincide with the phenomenon described by *Ayşegül Kibaroğlu* in this volume as the comprehensive security approach, namely the horizontal and vertical broadening of security studies, which creates a natural synergy between security studies and hydropolitics.

Nevertheless, mainstream security studies do not necessarily provide a sufficient vocabulary for studying transboundary water relations, and overemphasizing security could lead to incomplete results. One crucial point is put forward by *Ali Oguz Dirioz* when he describes that most scholars identify the state or society as the subject of security, which results in an anthropocentric viewpoint. A more integrative approach broadens the notion of water security beyond humankind, which could lead to novel research that would be partly incompatible with traditional security studies.

Moreover, as Kibaroglu's chapter shows, this tendency to interpret water relations in the realm of security studies has resulted in overemphasizing water disputes at the expense of cooperation. In a similar fashion, *Atal Ahmadzai*'s study juxtaposes the theoretical narratives of water wars and cooperation in water politics, showing that both logics can work well in their own theoretical contexts. Nevertheless, in practice, neither framework shows exclusive interpretative value for transboundary water issues, and arguably, almost all cases of transnational relations shaped by water show a mixture of cooperation and conflict with fluid intensity.

This has also been the case in the Middle East. As Dirioz shows, the fragility of political structures and relations, accompanied by the scarcity of water relations, has created an atmosphere dominated by conflictual interactions. That being said, cooperation does surface in various instances, and in the review of the history of interstate collaboration, Dirioz highlights how different states (namely Türkiye and Israel) and actors (including investors or the general public) can play a positive role.

Securitization and de-securitization have also been popular terms among scholars investigating transboundary water politics from a security perspective. The authors of this volume suggest that there is nothing automatic about this process. Depending on their geopolitical position, their relation with other stakeholders, their identity, and even their domestic politics, securitization can take place, although it does not necessarily take place, even if states have to suffer water scarcity. Additionally, water cooperation can actually help stakeholders de-escalate relations or even de-securitize other sectors through the experience and trust built up in managing transboundary water issues. The role of bottom-up initiatives and non-state actors should again be emphasized in creating win–win scenarios.

With all of these observations, several authors of this book also contribute to the evolving literature on a more coherent discussion of water diplomacy. One of the main results of our research in this regard is that it is not enough to avoid the conflict-driven narrative of traditional security studies, but investigating the nature of water diplomacy and the collaborative or hostile interactions between stakeholders of transboundary water management is key in each case. This assessment is in line with the suggestions to rethink the traditional framework of transboundary water interactions that is shaped by the dual nature of conflict and cooperation (Grünwald, Feng and Wang 2020).

In this manner, the authors of this book highlight different variables that can help explain the dynamics of complex interactions in various instances, building on the available literature (Offutt 2022). While putting the emphasis on individuals' constructs and trust, *Bota Sharipova* also lists other factors, including national interest, power accumulation, regime-building, and the existing institutional framework. On the other hand, *Péter Marton* interprets the interactions between upstream and downstream countries in the field of water-related public health challenges in light of the structural disparity between them. He argues that in general, when the upstream country is not exposed to health-related threats, the imbalance between the two sides hampers mutually beneficial cooperation. In this case, extra-sectoral incentives are needed to build up a collaborative environment. Interestingly, an intriguing debate could be spurred between Marton's arguments and those of Kibaroglu's regarding the ability and willingness of upstream and downstream countries to cooperate with each other.

Connecting the notions of water diplomacy and multistakeholderism, it is important to include non-governmental actors in the investigation of transnational water cooperation and dialogue. Some of these processes would be connected to state-actors (including Track-II diplomacy), as well as supra-state and sub-state actors. Several case studies presented in the book show the influential role and agency played by corporations, joint committees, consumers, or institutions of various level of authority and power in managing transboundary water politics.

Interesting dynamics of conflict and cooperation are also identified by *Zoltán Vörös* in connection with his analysis of China as a hydro-hegemon. Even in the case of severe disparity, states do cooperate with each other in a manner that serves the interests of the one in a hegemonic position. Power, he argues, is not necessarily exercised in a coercive manner. From this perspective, expressions such as leadership or dominance could be more useful in describing the nature of interactions in transboundary water politics as well.

14.7 Practical Implications

Most of these conclusions could be useful for scholars of International Relations and transboundary water politics in general, as well as those interested in the specific case studies the authors touch upon. Geographically, the chosen cases cover various regions and waters, including the Middle East, Europe, and the Pacific Ocean.

While the primary aim of our project was to contribute to the literature of IR theories, TWP, and regional and local studies, the conclusions of the various authors could also add value for practitioners who engage with transboundary water management. *Sameh Al-Muqdadi*'s study highlights the importance of slowing down dam constructions and advocates for the creation of a set of criteria (i.e. the "dam index") to be applied by the UN or other international institutions when evaluating a proposed construction project in the planning phase. *Máté Szalai* identifies the key circumstances that hamper cooperation and limit its effectiveness between small island developing states, including a lack of coordination on implementation or the tendency to create a novel institutional framework instead of focusing on the existing ones. These issues could be addressed by multiple stakeholders, paving the way for more effective and successful collaboration between states, international organizations, and various sub-national actors. Furthermore, *Najibullah Loodin* and his colleagues elaborate on why and how data sharing mechanisms can be developed in order to reach a more cooperative atmosphere in transboundary water management issues. Their approach is not only useful in terms of giving recommendations but also in the interpretation of the lack of progress in data sharing even when moving forward would be in the rational self-interest of all participants.

Other studies in the volume also give general or specific recommendations. *Péter Marton* advises downstream countries to introduce extra-sectoral benefits or compensations (for example, in the fields of trade or alliance policy) in order to make cooperation with upstream countries possible and equitable. Marton builds his argument on the analysis of the public health sector and the Danube river basin, but his recommendation could be useful in other fields as well. *Ali Oguz Dirioz* argues that in order to revive cooperation between Türkiye and Israel, increased inclusion of civil society and private business is needed, especially in the framework of the development of a more circular economy in the region.

This notion also connects our volume to the evolving agenda of the green transition both in Europe and around the world. As more and more actors dedicate their attention, resources, and political agency to building a green economy locally, regionally, and globally, water politics will play a key role not only in achieving sustainability but also in respecting the norms of justice, a peaceful resolution of conflicts, or mutually beneficial cooperation during the process. Without these viewpoints and without further research in water politics in the twenty-first century, green transition should not be considered as successfully implemented.

References

Cairns J (2001) Ethics in science and environmental politics: issues for interdisciplinary teams. Ethics Sci Environ Polit 1(1):51–56

Furlong K (2006) Hidden theories, troubled waters: International Relations, the "territorial trap", and the Southern African Development Community's transboundary waters. Polit Geogr 25(4):438–458. https://doi.org/10.1016/j.polgeo.2005.12.008

Grandi M (2020) Hydropolitics. Oxford Research Encyclopedia of environmental science. https://oxfordre.com/environmentalscience/display/10.1093/acrefore/9780199389414.001.0001/acrefore-9780199389414-e-644;jsessionid=0C665033E8516A26BFE5125F6EACC4F2. Accessed 7 February 2023

Grünwald R, Feng Y, Wang W (2020) Reconceptualization of the transboundary water interaction nexus (TWINS): approaches, opportunities, and challenges. Water Int 45(5):458–478. https://doi.org/10.1080/02508060.2020.1780854

Julien F (2012) Hydropolitics is what societies make of it (or why we need a constructivist approach to the geopolitics of water). Int J Sustain Soc 4(1–2):45–71. https://doi.org/10.1504/IJSSOC.2012.044665

Lake DA (2013) Theory is dead, long live theory: the end of the great debate and the rise of eclecticism in International Relations. Eur J Int Rel 19(3):567–587. https://doi.org/10.1177/1354066113494330

Marton P, Szálkai K (2017) Against the current: deconstructing the upstream/downstream binary in Hydropolitical Security Complexes. New 25(3):15–42
Offutt A (2022) Mixing waters: stakeholder influence in transboundary water conflict and cooperation. Water Int 47(4):583–609. https://doi.org/10.1080/02508060.2022.2059322
Raymond M, DeNardis L (2015) Multistakeholderism: anatomy of an inchoate global institution. Int Theory 7(3):572–616. https://doi.org/10.1017/S1752971915000081
Warner JF, Zeitoun M (2008) International Relations theory and water do mix: a response to Furlong's troubled waters, hydro-hegemony, and international water relations. Polit Geogr 27(7):802–810. https://doi.org/10.1016/j.polgeo.2008.08.006

SPRINGER NATURE

GPSR Compliance

The European Union's (EU) General Product Safety Regulation (GPSR) is a set of rules that requires consumer products to be safe and our obligations to ensure this.

If you have any concerns about our products, you can contact us on ProductSafety@springernature.com

In case Publisher is established outside the EU, the EU authorized representative is:

Springer Nature Customer Service Center GmbH
Europaplatz 3
69115 Heidelberg, Germany

The manufacturer's authorised representative in the EU is Springer Nature Customer Service Centre GmbH, Europaplatz 3, 69115 Heidelberg, Germany. If you have any concerns regarding our products, please contact ProductSafety@springernature.com

Printed and bound by CPI Group (UK) Ltd, Croydon, CR0 4YY
01/10/2026
02225892-0003